RED WATER, BLACK GOLD

RED WATER, BLACK GOLD

The Canadian River in Western Texas,
1920–1999

MARGARET A. BICKERS

Texas State Historical Association
Denton

©2014 Texas State Historical Association.

Library of Congress Cataloging-in-Publication Data

Bickers, Margaret A., author.
Red water, black gold: the Canadian River in western Texas, 1920–1999 / by Margaret
A. Bickers.
Includes bibliographical references.
ISBN 978-1-62511-002-2
1. Canadian River—Water rights—History—20th century. 2. Petroleum industry and
trade—Texas—Texas Panhandle—Water-supply—History—20th century. 3. Water
rights—Texas—Texas Panhandle—History—20th century. I. Title.
F392.C27B53 2014
976.4'8 —dc23
                                                                      2014030646
Design by David Timmons.

CONTENTS

ACKNOWLEDGMENTS

Although writers sometimes appear to work in a vacuum, our tomes and folios would not be possible without the assistance of a large number of people and institutions who encourage, assist, push, and sometimes kick us toward finishing our work. I give special thanks to the following groups and individuals.

This work would not have been possible without the very, very generous support of the Excellence in West Texas History Fellowship program. Administered through Angelo State University's West Texas History Collection, the West Texas Historical Association, and Texas Tech University, the fellowship encourages scholarship in all aspects of the history of Texas west of Abilene. Members of the board of directors of the organization provided advice, suggestions, and valuable critiques of this work while it was in progress as well as providing financial support.

Randy Vance, Tai Kreidler, and the staff of the Southwest Collection at Texas Tech University, including their student assistants and runners, provided a great deal of advice and made my research as pleasant as possible. Janet Neubarger directed me to the George Mahon papers and opened up a true treasury of information without which this book would remain incomplete.

Warren Stricker at the Panhandle Plains History Museum's archive provided encouragement and seemingly unending assistance (at least it probably seemed unending to him) during my years of digging in the archive.

The staff at the New Mexico State Record Center and Archive helped me wade through the state engineer's records and other related documents, even braving flaming potatoes in order to help researchers in need.

The librarian of the New Mexico State Engineer's Office graciously and cheerfully provided access to documents and publications within the office's archives despite being "under the weather."

The staff of the National Archives and Record Center–Denver pulled material relating to the Canadian River Project and suggested other possible locations for additional material to supplement their holdings.

The staff at the Amarillo Public Libraries central and southwest branches ordered obscure interlibrary loan books as well as providing access to the Bush/FitzSimmons Special Collection and the Genealogy Collection.

The Hutchinson County Museum and Hutchinson County Public Library, both repositories of critical material not available elsewhere, generously provided assistance and access to a wandering researcher.

The City of Lubbock's Water Department allowed access to their clippings and historical files.

Kent Satterwhite, general manager of the Canadian River Municipal Water Authority (CRMWA), and Tammy Hamby, the able and most generous office manager, opened their records to me as well as allowing use of CRMWA's lovely meeting room. Few archives provide such a magnificent view of deer, bald eagles, and winter skies or such complete access to their files.

Melissa E. Martin of the Missoula office of the Bureau of Reclamation and her assistants provided scans of the most pertinent parts of the bureau's last reports about the Canadian River Project.

Donna DeRight, city secretary for the City of Amarillo, provided access to the Amarillo City Commission minutes.

The Texas Natural Resources Commission made available aerial photos of the Canadian River.

Mike Harter once again provided ideas and encouragement as well as cautions as I waded into the second volume of the river's tale.

And last but certainly not least: Robert and Elisa Bickers provided reminders of life in the real world and that yes, there is such a thing. Peter and Gayle Bickers suffered through my discoveries (again), asked questions, pointed out authorial assumptions, and read the draft manuscript in its entirety as content and copy editors, and still like me.

All errors belong to the author.

To the storytellers and the story catchers who listened and wrote so that the rest of us could know.

PREFACE

This is a book for both scholars and interested lay readers. For that reason, the author has attempted to keep jargon and technical terms to the minimum necessary for clarity, and Appendix A includes a terms list.

Certain phrases and omissions are worth noting. In place of the more common "Euro-American" used to describe people of western European ancestry living in the United States, this work uses "Anglo American" or "Anglo Texan." The major role played by New Mexicans of Spanish descent, the Hispanos, in early Canadian River history means that "Euro-American" could be confusing. Spanish land and water use traditions differed from the English common laws introduced from the eastern United States into Texas. "Anglo Texan" better describes the people and the culture that came to dominate the High and South Plains economy in the twentieth century.

Readers interested in traditionally understudied groups in the High Plains will be disappointed in this work. Women, African Americans, and Latino/a residents of the plains contributed to the culture and economy of the region, especially in the cotton lands that extend from Lubbock south, and their struggles and stories are gaining more attention from historians. Very few Latinos/as, African Americans, or women achieved positions of political power in the region during this period, and this story focuses on the river and those politicians and regional leaders who drove the decision making. Indeed, African Americans and Latinos were systematically excluded from the political process until the 1960s. Poorer Anglos, along with Latino/as and African Americans, remain in the background of this chapter of the river's story.

Geography also requires some clarification. Those readers looking at a map of the area will note that everything north of the Red River resembles the handle of a pot or pan and might assume that "Panhandle" includes the entire region. That is not the case. The twenty-six most northern counties in Texas are the Panhandle or High Plains. From roughly Plainview (Tule Canyon) south to the end of the Llano Estacado is the South Plains. The Panhandle's economy centers on ranching and wheat, oil, and natural gas, while cotton dominates the sandier South Plains, with some oil production in the western and southernmost counties. Panhandle residents tend to look towards Amarillo as their market and shipping center, while South Plains dwellers favor Lubbock, "the Hub City of the South Plains."

A glance at the map also reveals an artificial limitation that this book places on the Canadian River itself. Eastern New Mexico remained part of the Canadian watershed during the period covered by this work. However, this book focuses on the Texas portion of the stream in Oldham, Potter, Hutchinson, and Moore Counties, and later on those South Plains towns that tapped the river for municipal use. As a result "western Canadian Valley" or "western Breaks" refers to the Oldham County stretches of the river unless otherwise specified. "River" and "stream" are used interchangeably, as they are in hydrologic works.

The term "drought" deserves some consideration. Drought means a shortage, usually of water. Agronomists, meteorologists, and geographers describe different types of drought. Meteorological drought refers to a period of below-average precipitation generally lasting at least three months, with average referring to the past thirty years' precipitation measurements. Agricultural drought implies that there was not enough moisture during the growing season; average rainfall may have fallen for the year, but not at the right time for local crops. Drought has social and economic aspects as well, such as causing population movement, including periodic regional depopulation. Businesses can be affected by drought as water-related recreation declines or farmers and ranchers spend less money because of their reduced incomes. Readers will be able to tell by the context of the term which form of drought is used in this book, and meteorological drought (or "drouth") is the most common.[1]

Readers will note that the author does not use the term "nature" to describe the nonhuman environment. The meaning of the word "nature" has changed greatly over time and means different things to different readers and commentators. As a result, "environment" and "physical environment" are the terms of choice for this book for the sake of clarity, and refer to those physical surroundings that were not created by humans. "Legal environment" describes the culture of laws, contracts, and decrees found in Texas and the larger United States during the period from 1930 to the present.[2]

Conversions to metric measures are included in the text except for water volumes such as acre-feet, cubic-feet-per-second and gallons-per-day. An acre-foot is the amount of water needed to cover one acre of ground one foot deep, or 325,851 U.S. gallons (1,233,482 liters). Water flow is measured in cubic feet per second (cfs), and one cfs equals 28.32 liters per second. The cities measured their wells in gallons per day (gpd) and priced water by cost per one thousand gallons. Appendix B includes a conversion table for those wishing to refer to it.

| DALLAM | Stratford | HANSFORD | Perryton | LIPSCOMB |
| Dalhart | SHERMAN | Spearman | OCHILTREE | Lipscomb |

County map of the Texas Panhandle and South Plains:

- DALLAM — Dalhart
- Stratford — SHERMAN
- HANSFORD — Spearman
- Perryton — OCHILTREE
- LIPSCOMB — Lipscomb
- HARTLEY — Channing
- Dumas — MOORE
- HUTCH-INSON — Stinnett
- ROBERTS — Miami
- Canadian — HEMPHILL
- OLDHAM — Canadian River — Vega
- POTTER — Amarillo
- CARSON — Panhandle
- Pampa — GRAY
- Wheeler — WHEELER
- DEAF SMITH — Hereford
- Canyon — RANDALL
- Claude — ARM-STRONG
- DONLEY — Clarendon
- COLLINGS-WORTH — Wellington
- PARMER — Farwell
- Dimmitt — CASTRO
- Tulia — SWISHER
- BRISCOE — Silverton
- Memphis — HALL
- CHILD-RESS — Childress
- Muleshoe — BAILEY
- LAMB — Littlefield
- Plainview — HALE
- FLOYD — Floydada
- MOTLEY — Matador
- COTTLE — Paducah
- Morton — COCH-RAN
- HOCKLEY — Levelland
- Lubbock — LUBBOCK
- Crosbyton — CROSBY
- DICKENS — Dickens
- Guthrie — KING
- Plains — YOAKUM
- TERRY — Brownfield
- LYNN — Tahoka
- Post — GARZA
- Jayton — KENT
- STONE-WALL — Aspermont
- GAINES — Seminole
- Lamesa — DAWSON
- Gail — BORDEN
- SCURRY — Snyder
- FISHER — Roby

Counties of the Panhandle and South Plains of Texas

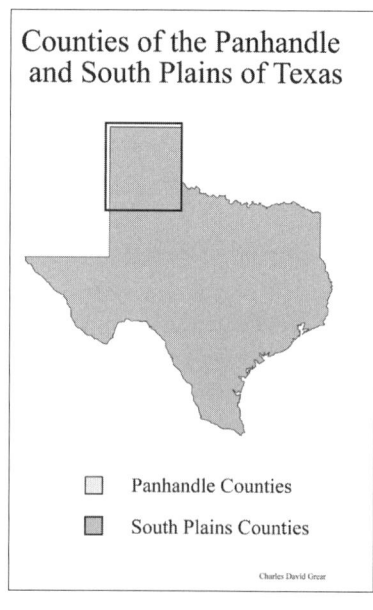

☐ Panhandle Counties

■ South Plains Counties

A Typically Atypical Stream

Than's it? This damp patch is the Canadian River?

Although it flows through a valley cut several hundred feet into the surrounding Llano Estacado plateau, and is spanned by bridges up to a mile long, the only hint of a real, live river hiding between the cottonwood trees and brushy salt cedar often comes from a small trickle or shallow stream. No wonder visitors new to the Canadian River Valley wonder exactly what happened to the river, assuming that the little ribbon of water under the highway bridge is, indeed, the Canadian. Those driving along the top of Sanford Dam glance over at the waters of Lake Meredith, note the apparent lack of river downstream, and perhaps assume that the dam explains the missing stream. And they probably wonder just how foolish people must be to build such a large dam on such a puny little trickle of river.

This book tells the story of that river in the twentieth century, explaining in part how the Canadian River came to look as it does and arguing that urban residents of the Panhandle of Texas chose to preserve their culture and economy by diverting most of the Canadian's waters to meet the needs of eleven cities and towns. The needs of the oil and gas industry encouraged this development, as did the rise of groundwater-based irrigation agriculture in the central and southern High Plains. In the process people separated the "wet" flowing river, the river in the Canadian Valley, from the "paper" river, the legal entity controlled by state and federal laws. At the end of the century a dispute over riverbanks and oil-lease royalties reconnected the wet and paper rivers downstream of Sanford Dam. Panhandle residents also created a new public space upstream of the dam, and then argued over the riverbed below the dam when physical changes to the wet river collided with the borders of the paper river. Texans argued with the federal government as well as with each other over paying for the capture and redirection of the river, eventually forcing the Bureau of Reclamation to sell the dam and its plumbing to High Plains residents for less than the initial contract price. At the end of the century the unpredictable Canadian turned the tables, confounding Panhandle and South Plains leaders despite their best efforts and forcing them to return to using groundwater and to reassert local control over the wet river in the process.

Three major themes emerge from the story of the Canadian and of the people around the river, all three related both by the river and by the idea of complexity. The unusual relationship between the federal government and local interests forms one theme, a relationship driven and in fact only permitted by the lack of federally owned land in Texas. This required federal, state, and local agencies to work with local landowners, negotiating more than would have been necessary for a dam and aqueduct on primarily public land. The second theme, as touched on in the previous paragraph, is that of legal complexity. Negotiations between states, between the cities and the federal government, and between the eleven cities themselves show yet again how tangled almost anything having to do with water is in the land west of the 100th meridian, whether that water flows through a valley or seeps underground. Oil and natural gas, the black gold of the twentieth century, form the third thread in the tale. They provided impetus for the development of a Canadian River dam in Texas by drawing industry and people to a poorly watered and otherwise sparsely populated place. Petroleum and gas production also consumed water, and the regulation of oil and gas production as compared to the lack of regulation of groundwater production provides an example of a road not taken. But the Canadian River never lacked for complexity even in the days before human manipulation of the stream became so obvious.

The Canadian River, once called the Rio Colorado or Rio Santa Magdalena, formed a vital connection between the Rocky Mountains and the Low Rolling Plains to the east of the Llano Estacado since at least the end of the last Ice Age. Paleo-Indians, Comanches and Apaches, Hispano shepherds from New Mexico, Anglo Texan ranchers, and water-craving townsfolk from hundreds of miles away used the river valley and its resources in order to sustain their cultures. For some the river provided spiritual benefits as well as physical necessities. Anglo Texans viewed the river's waters as a commodity to be conserved for the best use possible— slaking the thirst of town dwellers and providing entertainment through recreation. To this end Panhandle residents planned, labored, lobbied, argued, and pleaded until they got what they wanted.

This book tells the river's story in a roughly chronologic fashion, with a few slight diversions. The introduction of the book outlines the early history of the Canadian River and the environment through which it flowed prior to the 1920s. This section also explains some terms the author has chosen to use. The following chapters continue the river's story in detail through 1999. Although generally chronologic by decade and major event, two chapters step aside to explore the development of water law and of "paper" rivers and to look at life within the river valley shortly before the construction of Sanford Dam. Appendix A defines terms and acronyms in order of appearance in the main text, while Appendix B provides a conversion table and describes the methodology used for the climatic analysis of the region during the twentieth century. Appendix C outlines the "history of the history" of western rivers and of the Canadian's watershed while offering further titles for interested readers.

• • •

The Canadian River begins in the mountains of Colorado, just north of the New Mexico border. It flows south past Raton, New Mexico, on the foreland of the Sangre de Cristo Mountains, where a few small irrigation projects tap its waters. Then the river twists and cuts into a gorge almost a thousand feet (300 m) deep. The gorge first formed hundreds of thousands of years ago when the volcanic energies of the Jemez Lineament pushed up from below, raising the land around the river. To the north and east, volcanoes such as Sierra Grande and Capulin erupted, but here the Canadian simply wore through the layers of older volcanic rocks and younger lava flows. At the junction of the Conchas and Canadian Rivers, the stream makes a ninety-degree turn and begins flowing east–northeast, across the eroded, mesa-dotted plains of eastern New Mexico and into Texas.[1]

Once in Texas, the river again sinks below the level of the surrounding landscape. It runs east over a sandy bed, jogs slightly north roughly 120 miles (190 km) after entering the Llano Estacado Plateau, then returns to its eastward path. The broad, rough double valley of the Canadian grew slowly as surface and groundwater eroded the layers of sandstone and salt that formed the foundation of the huge, generally flat surface of the Llano Estacado, the "staked plains" of Texas and eastern New Mexico. The "Breaks" (as they are known) developed from a series of sinkholes and collapses caused by groundwater dissolving the layers of salt

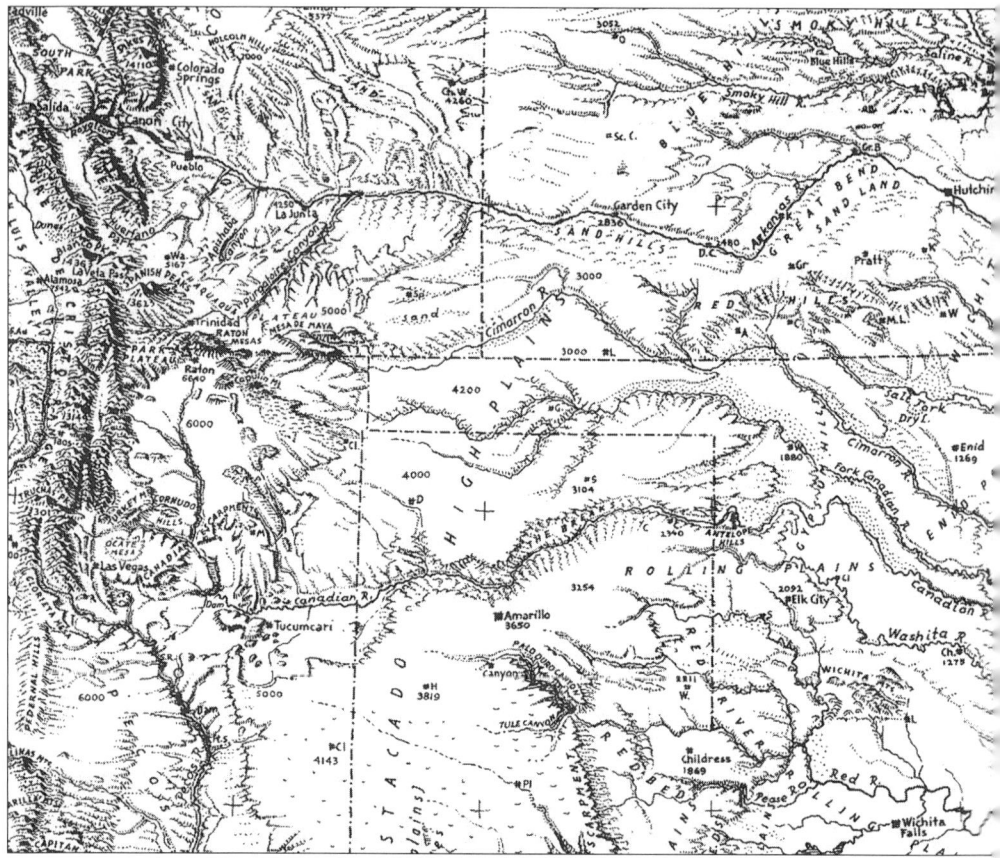

Figure 1: The Canadian and the Llano. Detail from Landforms of the United States, *6th Revision, 1957 (C) Erwin Raisz. Used with permission of the Raisz estate. All rights reserved.*

and gypsum laid down by ancient seas. Over millions of years rainwater and snowmelt trickled down through loose layers of gravel and cracked sandstone into the buried salts, joining older water already trapped in the rocks as it percolated southeastward through the ground. The water ate away the salt and gypsum, causing the ground above the layers to collapse and form sinkholes and giving the waters of the Canadian a distinctly brackish taste.[2] Over time the sinkholes grew and connected. Water from streams sought the low points on the plains and eroded the land further, slowly chewing westward from the damper eastern edge of the plateau until the present double valley lay open to the sun. From a distance north or south of the Breaks the plains seem to extend forever, aside from a

RED WATER, BLACK GOLD

faint blue line on the horizon. Approaching closer to the river, the land becomes sloping and rolling and streams cut deeper channels—this is the outer edge of the Breaks. Soon, the broad, deep main outer valley becomes visible, sprawling up to 600 feet (183 m) deep and 20 miles (32 km) north to south. Side canyons feeding into the main river channel cut into the Ogallala Formation, releasing sweet groundwater and creating springs that in turn supply streams. Deeper within the Breaks, the river flows through the inner valley, what most people would think of as the proper river valley.[3]

Although separate from the Canadian River, the Ogallala Formation and the water trapped within its gravels and sand helped make and feed the river. The mountains in what are now New Mexico and Colorado provided the material for the rock layer that formed during a time of wetter climate. Streams and rivers carried bits of rock (sediment) away from the young peaks and left them on the plains as far to the east as central Oklahoma and Kansas. When the rivers filled their channels with this sediment, the water changed location, covering more land with bits of mountain. Over time the climate grew drier, weakening the streams so that they no longer reached the plains, and erosion replaced deposition. To the west, the Pecos River cut its way northward, catching the older rivers flowing off the Rockies and in the process separating the Llano Estacado from the Rocky Mountain foreland. At the same time the eastern edges of the Ogallala eroded away. A cap of harder material protected the water-bearing layer on the Llano Estacado, creating a plateau rising more than 3,500 feet (1,067 meters) above sea level. Where erosion cut into the Ogallala, the trapped water emerged as springs, feeding some small lakes and contributing to streams that fed the Canadian, Red, and Brazos Rivers. This "fossil" water recharged very slowly from rainfall and snowmelt on the plains and provided much of the cold-season flow of the Canadian.[4]

The Breaks around the river provided a special habitat for animals and plants in the region. Above the Breaks, the semiarid climate of near-constant wind, bright sun, and low rainfall contributed to the dominance of short grass steppes, grazed by bison, pronghorn antelope, jackrabbits, and prairie dogs. Rainwater lakes, called playas, provided some water in wet years or during the spring and early fall rainy seasons, but the shallow basins often went dry in the late summer heat, their water stolen by hot southwesterly winds. The Breaks provided shelter and water year round, making a habitat connection between the uplands to the west and the well-watered Low Rolling Plains to the east. Here grew hackberry and cottonwood trees, some mesquite and willows and varieties of oak. Along the river and streams, tule reeds throve in marshy wetlands, while thickets of wild plums and large vines bearing wild grapes could be found on the riverbanks. Tall grasses such as big bluestem and Indian grass provided food and shelter close to the streams, while mixed and short grasses such as gramas, little bluestem, western wheatgrass, and others grew on the broken ground farther from the river's edges. Some juniper and mesquite, as well as arid species such as cacti and yuccas, clung to south-facing slopes where evaporation dried the soil more than in other parts of the valley. This variety of plants encouraged a variety of animals to make the Breaks their home or seasonal refuge.[5]

The Breaks are an "edge habitat" where the semiarid steppe meets the Low Rolling Plains. Imagine leaning back against the rough trunk of a cottonwood tree on a soft spring morning in 1800, just before sunrise. A faint breeze moves the leaves, and a splash sounds over the soft rustle, then another splash and plop. Beavers drag newly cut tree limbs into a stream-fed pond in order to repair their dam. Heavy rain from a thunderstorm three days before caused a flash flood, and the spring-fed stream rose quickly, tearing out the edge of the dam. Other motion down by the river signals the arrival of white-tailed deer. They emerge from the brush edging this part of the riverbanks and the small herd drinks from the river, wary for the presence of bobcats and mountain lions. Black bear also stalk the Breaks, and an occasional elk (wapiti) wanders down from the highlands to the west. Something spooks the deer and they flee back into cover, scattering a sleepy covey of quail that flush with a "whirrrr." Wild turkeys stalk and scratch under cover of the trees. It is the wrong time of year for bison to visit the valley except in passing. Their season is winter, when they seek out the shelter from the wind and snow. Now they migrate across the steppes above, coming through the valley only en route elsewhere. Water is plentiful now, and the bison prefer the open plains, where the wind carries warnings of the wolves that stalk the great herds. Another predator also haunts the uplands and the Breaks as well: humans.[6]

• • •

The first known human traces in the Breaks date to the end of the last Ice Age. People of the Clovis culture followed the megafauna, the mammoths, giant bison, camels, and other edible game across the plains. It is probable that while they hunted the diminishing numbers of giant mammals, the Clovis people also made use of the plants and fish of the plains to a great extent. (There may have been people on the plains before the Clovis culture, but the archaeological record for this region remains blank prior to roughly 12,000 BCE.) After Clovis came the Folsom people, also big game hunters who followed the giant bison, *Bison priscus* and *B. antiquus*. As with Clovis, little remains of the Folsom culture besides bones and stone tools, notably the beautifully chipped and fluted lance-blade-like points first found in the volcanic area of the eastern New Mexican plains.[7]

The available evidence suggests that it was not until the 600s or 700s CE that people lived year round in the central region of the Canadian River's watershed. To the east, in the Mississippi River Valley and watershed, the great mound-builder culture arose and prospered, spreading elements of its religion and manufactures as far as southwestern Kansas. The Colorado Plateau to the west of the High Plains proved congenial to the Mogollon and Anasazi peoples, at least until drought and probable overuse of limited resources forced the people of Chaco Canyon and Mesa Verde to move east into the Rio Grande watershed. Trade goods crossed the High Plains as seashells from the Pacific coast passed from hand to hand until they reached the Mississippi Valley, while native copper from the Great Lakes area was carried west and south. Pottery found in the Canadian watershed indicates that exchanges between the Plains dwellers and the Rio Grande residents predated the last prehistoric culture in the Canadian Valley, the Antelope Creek people.[8]

The Antelope Creek people were the last to live permanently in the Canadian Valley until Hispano pioneers from New Mexico ventured out of the Rio Grande watershed in the 1870s. The Antelope Creek people built blocks of dwellings within the Canadian Breaks that resembled the pueblos of the Rio Grande Valley, but that reached only one story. Instead of adobe, flat rock-slabs formed the base of the "pueblo" walls. The settlement dwellers farmed the Breaks, growing maize, beans, squash, and other crops with the aid of irrigation water from the springs and seeps. Breaks residents also made use of wild plants, including the plums growing along the river, cattails and arrowhead plant, and grapes. The Antelope Creek people hunted the Breaks and the plains, killing bison, deer, and smaller game. They traded east and west, using the Canadian Valley as a highway. And then, before 1500 CE, they departed the region.[9]

Archaeologists remain uncertain of what caused the Antelope Creek people to depart the Breaks. Climate reconstructions suggest that a drought enveloped the Southwest and High Plains, causing the bison to move east and north and reducing the rainfall and river flow below what would sustain the people's crops. This was also the time that the first of the Athapaskan peoples, the ancestors of the Navajos and Apaches, moved south along the Rocky Mountains, and it could be that in the contest for sparse resources, the proto-Apaches drove out the Canadian Valley farmers. More likely, a combination of drought and overuse of the remaining resources (wood, wild plants, game) eventually forced the Antelope Creek people to abandon their towns and fields, moving east to the wetter Low Rolling Plains and tall-grass plains, or west into the Rio Grande Valley. The physical environment of the High Plains did not favor long-term permanent settlement, even in the well-watered Breaks.[10]

The Apaches and Comanches who moved onto the High Plains took advantage of the Breaks' resources. Both groups hunted the bison of the plains and the deer, bear, elk, and other game that sheltered in the Breaks. They also made use of the wood and water and sheltered valleys of the larger Breaks, especially in winter and the height of dry summer. Like the Antelope Creek people, the Apaches raised some maize, beans, and other crops, but this proved fatal. Once the Comanches reached the High Plains, they caught the Apaches at their fields and over time drove them south and west. During the nearly two centuries of Comanche dominance of the Canadian watershed, they never took up farming or built fixed settlements. Instead the Comanche bands moved their horse herds and camps as necessary to find greener pastures or bison herds or to meet with trading partners to the east and west. Through hunting, trading, raiding, and horse breeding, the Comanches developed a cultural dominance that forced even imperial Spain and France to treat them with care.[11]

However, the Comanches needed access to outside resources in order to prosper on the High Plains. They exchanged horses and bison products for maize (and later for wheat bread), cloth, decorative items, obsidian, and slaves from the Rio Grande Valley at trade fairs in places such as Taos Pueblo. The Comanches also traded eastward with the Wichitas and other residents of the Low Rolling Plains, and sold horses north to the Cheyennes and similar peoples who were unable to sustain large horse herds over the winter. When Europeans arrived within the Comanche trade networks, metal goods and glass items and eventually firearms joined the list of exchanges. During drought years or if the bison "harvest" failed, the Comanches of the High Plains experienced difficulties because of their need for imported carbohydrates. This dependence was one sign of a problem that eventually caused great difficulties for the Comanches.[12]

It appears likely that the Comanches overstressed their physical environment in the nineteenth century. After acquiring horses and moving onto the Great Plains in the early 1700s, the Comanches raised ever-larger numbers of the equines. They practiced selective breeding to a point, but also viewed horses as symbols and evidence of wealth and status. The needs of their herds forced the Comanches to remain nomadic, moving every time the grass gave out. In winter, if the grass failed or snows buried it too deeply, the Comanches cut cottonwood bark for fodder. During the long drought of the 1850s and 1860s it appears that the Comanches deforested parts of the Texas Canadian River Valley while attempting to feed their herds. Their increasing need for bison in order to meet the demand of Anglo and Hispano traders for meat and hides, when combined with the destruction of the northern bison herds, put great stress on the Comanches' culture even before commercial hunters began attacking the southern bison. Comanche horses also competed with the bison for food, notably in winter, when the bison sought the shelter of the Breaks and of other canyons and river valleys. When the bison population collapsed, the resource shortage pushed the Comanches into an ecological corner. Some bands tried to replace wild bovines with stolen domestic cattle, but the arrival of the U.S. Army during the Red River War of 1874–75 terminated what may have already been a tottering lifeway. The resulting depopulation of the Canadian Breaks opened the way for Hispanos to move onto the plains.[13]

Hispano New Mexicans followed the Canadian out of the Rio Grande watershed and into the plains even before the American defeat of the Comanches. Spaniards and (after 1820) Mexicans ventured onto the plains as traders, herders, and soldiers, or were brought onto the plains as Comanche captives. They occasionally married into Comanche bands or developed fictive kinship relations with some Comanches in order to trade and hunt the bison plains. After 1848, as the demand for New Mexican sheep soared in the wake of the gold rushes in California and Colorado, the pastures of the Rio Grande basin grew overcrowded, forcing braver souls to look eastward. Carefully and tentatively at first, Hispano settlers moved east of the Sangre de Cristo Mountains along the network of streams that fed the Canadian, Pecos, and Dry Cimarron Rivers. Soon a few settlements developed first at the foot of the mountains and then gradually toward the High Plains.[14]

As soon as the "grazing niche" on the Canadian River opened up, the Hispanos moved into it. Family groups, including the Romero-Duran clan that settled the Canadian Valley in the western Texas Panhandle and the Gallegos family of Ute Creek north of the Canadian, established settlements called *plazas* along the main streams. Large flocks of sheep grazed their way onto the plains and back to the plazas or even to the mountains in a type of transhumance, or seasonal migration and grazing. The Hispano settlers practiced an extensive use of the Canadian's resources, tapping the river, the Breaks, and the uplands for different materials at various times of the year. They gathered fruit, fished or built fishponds, hunted the deer and smaller animals, grazed their flocks and grew maize, beans, squash, apples and peaches, and chiles and other vegetables. They also trapped beaver and muskrat when possible. Hispano sheep owners sold their sheep and fleeces in Las Vegas, New Mexico, or Dodge City, Kansas, as well as gathering and selling bison bones and in one case making and selling salt from an especially strong brine spring north of the river. The Hispano settlers were successful at first. How long they could have continued in the Canadian Valley remains unknown. Herds of cattle and Anglo Texan ranchers pushed into the Breaks not long after the Hispanos began building their homes.[15]

• • •

T. S. Bugbee founded the first ranch in the Canadian Breaks, but his operation was not alone for long. Located on Bugbee Creek, roughly 20 miles (35 k) east of Romero's plaza, the Bugbee Ranch marked the first Anglo cattle operation in the valley. Others followed very quickly, and soon unofficial claims filled in the Breaks. George Washington Littlefield founded the LIT Ranch between Bugbee's spread and Romero's plaza. To the west, businessman and federal contractor William McDole "W. M. D." Lee and a partner created the LS Ranch. The Hispano shepherds found themselves pushed back into New Mexico as Texas cattle replaced bison in the valley. With a few notable exceptions, these ranchers were not the rugged individuals of western legend but businessmen backed by financiers from Chicago, Kansas City, or even New York and London. The Canadian Valley formed a pocket of unused and thus far unclaimed resources where land remained available in large blocks for purchase or lease from the State of Texas. The open range closed quickly.[16]

It is worth noting that there was no federal land for homesteaders in the Texas Panhandle. The annexation treaty between the Republic of Texas and the United States stipulated that the new state would keep all its public lands, which would be held and disposed of for the good of all state residents. As a result, while the state encouraged homesteading and provided legal means for farmers to settle the land, the state also made much larger blocks of land for sale or lease to livestock raisers than did the federal government. The land in and around the Canadian River filled in quickly as a result, because control of the water meant control of the surrounding real estate. Across the border in New Mexico, the federal government did not permit land leasing or sales of more than 160 acres per individual, with some exceptions, as the ideals of the Homestead Act collided with the climatic realities of the West. This difference in land availability marked one of the first divisions of the Canadian watershed.[17]

Over the next decade the land ownership patterns in the Canadian Breaks developed in favor of the eastern-backed ranchers. From west to east, the valley filled with cattle, and fences closed the formerly open range. The westernmost property, the Capitol Land Syndicate, contained 250,000 acres (101,250 ha) of the High Plains and Breaks within 6,000 miles (9,650 km) of barbed wire fence. Most people knew the ranch by its brand: the legendary XIT. The goal of the ranch's English owners was simple: raise cattle to make money until they could sell the land to home-steaders and farmers. To the south, W. M. D. Lee broke with his first business partner and sold the LS, creating in its place the LX Ranch to the east of Tascosa. The English-funded Prairie Land and Cattle Company bought out Littlefield's LIT only to discover that he did not own all the land he had claimed, forcing them to purchase and lease it twice, the second time from the State of Texas. Farther east on the southern edge of the Breaks, English and New York creditors took over the Francklyn Ranch to create the White Deer Land and Cattle Company, leasing some pastures to the Scottish-owned Matador and other ranches while making plans to sell the rest to homesteaders. Romero's plaza became the town of Tascosa, a market and shipping center hemmed in by the XIT and Lee's ranch. The cattle crash and die-offs of 1885–86 and 1887–88 shook the Panhandle ranchers further, leaving new owners scrambling to explain to creditors where their fabulous cattle fortunes had gone.[18]

Railroads arrived in the late 1880s, buoying land prices and terminat-ing the prospects of Tascosa and other bypassed towns. The Fort Worth and Denver Railway reached the central Panhandle in 1888, prompting the creation of the town of Amarillo near spring-fed Wild Horse Lake just south of the Breaks on land owned by the LX. The adjacent Frying Pan Ranch, between the LX and Tascosa, chipped in some land as well. Despite the fervent wishes and prayers of developers at Tascosa, the rail-road initially stopped south of the Canadian, later crossing the river at a new bridge upstream of the hamlet. Panhandle ranchers took advantage of their location north of the so-called dead-line for the dreaded Texas cattle fever, shipping cattle year round and selling their clean animals for higher prices than "Texas cattle" received. Soon cattle from south of the Cana-dian marched up the prairie and watered at Amarillo before going via rail to Fort Worth and Kansas City. Those XIT divisions located north of the river preferred to ship cattle to Chicago. A few farmers settled along the railroad in the southern and eastern parts of the Canadian and Red River watersheds, where rains fell more often and cattle remained sparse.[19]

Despite the best efforts of land promoters, farming in the Panhandle failed to boom until after the mid-1890s. Economic depressions and drought in the 1880s and early 1890s contributed to the slow expansion of farming. Because Texas owned all of its land, there were no "free" government homesteads available as there were in New Mexico, Kansas, and other western states. Many agriculturalists also remained skeptical that the High Plains could produce marketable crops aside from cattle, given the dry climate. The lack of railroads and of access to national and international markets further slowed settlement. Within the Breaks, the rugged land discouraged thoughts of farming, aside from some small, irrigated areas and experiments with fruit cultivation. Some boosters blamed the stockmen for discouraging settlement and even driving off smallholders as the English ranches southeast of the Panhandle were reputed to do, but the climate and topography did more to check the advance of the farming frontier into the Panhandle than did "big ranchers."[20]

Nevertheless, settlement continued despite drought and national economic depression. By 1907 railroads stretched across the Panhandle and into the southern plains, as well as pushing into eastern New Mexico for the first time. Revisions to the federal homestead laws encouraged people to settle in the New Mexican segments of the Canadian watershed, displacing many small Hispano settlements and sheepherders in the process. Developments in dry farming of wheat and the introduction of hard winter wheat varieties and various types of sorghums, along with the dissolution of the XIT and sale of the White Deer lands, brought farmers into the Canadian watershed. However, the Breaks remained ranch land. As the Capitol Syndicate broke up, the Matador Land and Cattle Company bought all of the XIT's Canadian River lands, creating the Alamositas Division of the Matador. Downstream, Amarillo businessman Lee Bivins bought the old LIT and LX, while others obtained more of the river valley. By the beginning of the First World War, Anglo Americans established an economic and cultural pattern that would remain in place for the next century, a pattern centered on sunlight, soil, and water via farming on the plains and ranching in the Breaks.[21]

The combination of year-round grazing and shifts in the rain and snowfall patterns of the region contributed to changes in the Canadian River during the years between the Red River War and World War I. Casimero Romero's son, José Ynocencio Romero, noted that cattle had broken down the formerly high and grassy banks of the Canadian near Tascosa. They also ate the grass off the lower banks, and Romero believed that as a result the river became wider, shallower, and more intermittent. It is probable that before the introduction of large numbers of windmills and water tanks, cattle and horses overgrazed the streams and playas as well as the Canadian's inner valley. However, a series of droughts followed by heavy summer storms caused erosion of the uplands that introduced more sediment into the river, causing some of the widening. Alterations in the pattern of rain and snowfall contributed as well. Instead of relatively consistent, year-round moisture, precipitation became more seasonal and concentrated in spring and summer. The pattern shift also accelerated erosion because winter-dry plants could not hold the soil when the first heavy rains came down. Deforestation of the valley further sped erosion by eliminating beaver, whose ponds and dams had served to slow the flow of water and to check soil losses.[22]

In 1917, the Canadian ran free. A few attempts to irrigate river flats upstream of the Canadian-Conchas confluence failed when high water ripped out the check dams and irrigation gates, adding insult to injury by submerging the irrigated farmland and drowning the crops. The Matador Alamositas division manager grew some irrigated alfalfa by tapping a stream and groundwater. Ranchers generally saw no point in wasting resources by trying to tame the silty, salty, and unpredictable Canadian. A drought such as that of 1903–04 might break with a record-setting flood, as hit the region in late September and early October 1904. A dozen souls perished, and bridges from New Mexico, Texas, and Oklahoma washed downstream on the river's churning flow. Drought gripped the region yet again in 1917–18, forcing people to abandon homesteads in eastern New Mexico, the western edge of the Panhandle, and even the Texas Trans-Pecos. The Canadian's flow dwindled and faded to the point that in some places people walked over the riverbed dry-footed. Then a local storm could cause a sudden rise that left deadly quicksand behind. The Canadian remained wild, a stream worthy of respect and caution.[23]

Chapter 1

OF AGRICULTURE AND OIL: NEW RICHES FROM THE GROUND

The petroleum industry roared into the Panhandle when the Masterson Number Two well gushed natural gas and oil from the Canadian River Breaks, opening the Dixon Creek field and ushering in the Panhandle's oil boom during the "Roaring Twenties." For farmers, the 1920s marked the onset of a depression as wartime prices fell, drought came and went, and tractors replaced hired hands. At the same time, gasoline-fueled tractors driven by "suitcase farmers" (people who spent only a few weeks a year working the land to raise wheat while keeping another job in Amarillo or another city) broke new ground for wheat in the far western Panhandle on the land surrounding the Canadian River Valley. While politicians argued over how best to deal with the growing "farm problem," some Panhandle landowners began earning more from leases and royalties than from crops and cattle. Within the Breaks, wooden derricks sprang up, marking the site of the Panhandle's newest industry, and towns such as Borger appeared as if overnight on the edge of the valley, booming and blowing almost as hard as the gas wells and gushers around them. The oil-boom growth brought water-supply problems to the cities of Amarillo and Lubbock and the new town of Borger, causing covetous eyes to look again toward the Canadian's "wasted" waters. At the same time, developments in Washington, D.C., and elsewhere began lapping the region like the first wash of waters from a rising tide.

Events far from the Breaks affected the Canadian River. Not immediately, of course, but the trends in national policy and regional economics that developed in the period between the First World War and the onset of the Great Depression would continue and intensify over the next eighty years. The region's economy depended on extracting solar energy and water from the ground in the form of grasses such as wheat and maize (corn), either on their own or after processing through cattle, and on the long-dead plants and microorganisms that formed petroleum deposits and natural gas. Selling and shipping those goods to buyers around the nation brought other goods and materials back to the plains. If the halls of the Department of Agriculture and the floor of Congress remained spatially remote from the Canadian watershed in 1920, the effects of decisions and actions taken there and in commodities markets in Chicago and London greatly affected the people living on the High Plains and South Plains.

For some observers of national agriculture in the 1920s, the agricultural depression of that decade came about as much from misuse of natural resources as from poor decisions by individual farmers and national policy makers. Commodity prices declined rapidly after the end of World War I, dragging many farmers deeper into a cycle of loans and debt even as other Americans prospered. In the Canadian River watershed, increased mechanization and technological advances proved to be a blessing for those in need of irrigation and a burden for farmers trying to pay off new tractors, combines, and other equipment. Federal attempts at assisting farmers provided short-term help of a sort but did not stop the long-term problem. As a result, conservationists such as Hugh Hammond Bennett looked at the developing situation and wondered if there was more going on than simply overexpansion of farms and an international agricultural trade imbalance.

The agricultural depression had begun in 1920 when prices for most commodities collapsed following the end of World War I. Farmers in the Texas Panhandle, South Plains, and elsewhere planted as many acres as possible in 1919 on the assumption that the high wartime prices would continue for at least another year. However, European agriculture recovered much more quickly than farmers in the United States had anticipated. Exports of farm goods declined, leading to large surpluses in the U.S. grain market. Bountiful harvests in Canada and Argentina, offered at lower prices than U.S. wheat growers asked, encouraged international buyers to purchase from elsewhere in the Western Hemisphere. At the same time, the federal government removed price controls and price floors because the end of the war meant the end of wartime needs. The change affected the prices of wheat, cotton, beef, corn—the Midwest's and High Plains' major crops—and farmers' incomes dropped below parity, where they would remain for the next two decades and more. Wheat that had been supported with a floor price of $2.00 per bushel during the war dropped to $1.09 between July and December 1920. Livestock prices also declined: cattle went from the 1917 high of $8.40 per 100 pounds (cwt) to $4.47 per cwt and hog prices tumbled from the early 1919 peak of $18.60/ cwt to $6.90. As happened after the Panic of 1893, Texas farmers and ranchers found themselves in dire financial straits, now exacerbated by new expenses for the equipment and the additional land bought or leased during the war to produce still more wheat, corn, and sorghum.[1]

While declining prices affected everyone, they hit those who had purchased more land and machinery the hardest. Tractors, even small ones, cost a great deal of money, and many farmers found themselves in debt just as their incomes began declining rapidly. As prices for agricultural products fell, many farmers tried to make up the difference by planting more acres. While this might work for an individual, when most farmers planted more ground, it simply increased supplies, depressing prices and continuing the cycle. Implement dealers still expected to be paid and now it took more and more wheat, cattle, or corn to cover the cost of the new tractor (and fuel, oil, spare parts). In a second blow, because soaring demand had driven up the price of farmland during the war, property taxes rose and remained high. Farmers were caught between falling prices for what they sold and steady prices for what they bought, and by 1921, their average purchasing power declined to two-thirds of what it had been during the "golden era" of 1910–19. The increased national standard of living and memories of the new amenities that farmers had purchased during the good years only added to their frustration and the sense that they were falling behind while running faster and faster. The low food costs came as a boon to city dwellers and those not directly involved in the agricultural economy, but the people producing that cheap food tried a number of different solutions to solve the problem and raise their incomes.[2]

Tools used to raise prices, or at least to stop the decline, included crop diversification, crop reduction, and requests for national assistance. Randall County, south of the Canadian Valley, had become something of a dairy center, supplying Amarillo and more distant markets with milk and cream. Poultry production also tripled, serving as a way to convert corn, sorghum, and wheat into a higher-value product. The continuing demand for wool encouraged some ranchers and farmers to turn to sheep raising, and the number of woolies climbed from 47,216 to 120,250 between 1919 and 1930. In contrast, the number of pigs sank along with pork prices. Farmers returned some land that had been broken for dry-land crops to grazing or simply abandoned the land as the rents and leases expired. Wheat acreage declined somewhat in the early 1920s in the High Plains as farmers retrenched. Some shifted to other, higher-profit crops as a drought in the mid-1920s further cut regional surpluses. Local banks foreclosed on a few overextended farmers. Other farmers turned their attention to Washington, D.C. In their opinion, the federal government had caused the problem with high price guarantees during the war; therefore, the federal government should help undo the damage that had been done.[3]

Individuals and agricultural groups such as the Farm Bureau, the Farmers Union, and the Grange proposed several solutions to the "farm problem." In Texas, wheat growers banded together as early as 1921, marketing their grain as a group or pool in an effort to increase prices. The Texas Wheat Growers Association enjoyed some success; in 1924 association members on average received $1.26 per bushel as compared to $1.10 for nonmembers. On the federal level, advocates for federal involvement argued that price floors to guarantee parity should be restored, at least temporarily. Others wanted a system of federally funded credit based on crops stored in government-owned storage facilities, in effect making the national government the country's main grain dealer and agricultural lender. There was also talk about co-ops and even of "farm holidays" as had been tried in the 1890s (and would be revived in the 1930s). Some of the ideas mooted traced their origins to the Farmers' Alliance and the Populist Party of the late 1880s and early 1890s. Almost all of the proposals envisioned a larger role for the federal government in agricultural operations. Farming was becoming a business much like steel manufacturing or banking, with its own lobbies (the Farm Bureau in particular) and political blocs.[4]

Some legislators attempted to ease farmers' financial problems. In 1921, U.S. representatives and senators from the Midwest and South formed an informal "farm bloc" to coordinate legislation that included increased money for loans, national oversight of livestock pricing through the 1921 Packers and Stockyards Act, and later a restriction on grain and livestock futures trading aimed at preventing excessive price fluctuations. Arthur Capper of Kansas and Andrew Volstead of Minnesota sponsored the Capper-Volstead Cooperative Marketing Act in 1922. They hoped that by allowing farm groups, such as the Texas Wheat Growers Association or a national cotton farmers group, to hold back their crops or to market their crops as a bloc without running into difficulty with the Clayton Antitrust Act, the bill would smooth the seasonal drop in prices that came every summer and fall when crops were harvested. However, despite these and other farm bills, prices seemed to stagnate, and the Farm Bureau and others wanted more.[5]

The McNary-Haugen bills, a series of measures introduced and vetoed in the later 1920s, show just how much more assistance some farmers and their advocates desired. Senator Charles McNary of Oregon and Representative Gilbert Haugen of Iowa proposed that the federal government establish a system of dual prices for agricultural commodities. Tariff walls would protect domestic prices while the federal government calculated if there was a crop surplus. If a domestic surplus existed, the government would sell that extra grain, sugar, or meat on the world market at world prices. This would keep domestic prices high enough to bring farm income back up to "parity." If the United States had a poor wheat harvest, the federal government would not sell any grain abroad and would allow foreign stocks onto the U.S. market, either at the same price as domestic supplies or taxed at a higher, still protective, rate. All this would be managed by a federal export corporation and funded by an "equalization fee," a tax on each bushel or unit (bale, cwt) of crop sold. In other words, the federal government would become the world's largest commodities dealer, and all farmers (and eventually consumers) would pay to make up the difference between the U.S. and world prices, should global values decline below domestic prices.[6]

When McNary and Haugen proposed their first version of the bills in 1924, Congress opposed it on grounds that they should not raise food prices simply to benefit one small (and shrinking) portion of the population. Undaunted, the pair reintroduced versions of the basic bill in 1927 and 1928; each time Congress passed the bills and each time President Calvin Coolidge vetoed it. "Silent Cal" argued that the bill included an illegal tax on farmers and that it favored grain producers over dairy farmers and other grain consumers. He also pointed out that higher prices would encourage more grain production, lowering prices again and exacerbating the current problem. President Coolidge preferred to see cooperative and orderly marketing efforts by pools and associations of farmers acting independently of the federal government, lest a new (and expensive) bureaucracy develop. However, following a further price collapse in 1929, a bill closely resembling parts of the McNary-Haugen plans passed. The Hoover Agricultural Marketing Act provided a $500 million loan to agricultural cooperatives so they could buy up the surplus and thus keep it off the market until prices rose again.[7]

Even as the bills wound their way through Washington, D.C., farmers in Texas and other places had not waited for the federal government. They took their own steps to buoy their finances and stability. For Texas Panhandle farmers, the later 1920s proved to be a period of diversification and gradual expansion westward as wheat prices slowly rebounded, then surged. After the initial pain of the price collapse and a severe drought in 1923–24, things began improving somewhat. Agents of the new state and federal Agricultural Extension Service offered advice on new crops and varieties of wheat, sorghum, and other grains, while the Future Farmers of America and Future Homemakers of America programs encouraged young people to stay in agriculture while learning the most scientific and current ways to farm and to run a farm household. As corporate owners broke up and sold the western Panhandle ranches, more, smaller farms appeared on the land. Many of these were stock farms, and the numbers of dairy cattle, sheep, and poultry increased. There were three times as many chickens reported in the agricultural census of 1930 as in 1920, and the number of sheep doubled, although the pig population declined. Rabbit plagues in 1924 and 1925 affected some farmers, but the region as a whole remained steady, if not exceedingly prosperous. The decade's rapidly increasing mechanization also accelerated crop diversification, especially in the South Plains.[8]

Two developments, the appearance in 1924 of the International Harvester Farmall—the first general-purpose, gasoline-powered farm tractor—and the introduction of the power takeoff unit, which allowed owners to use the tractor to run other pieces of equipment including augers and pumps, encouraged more and more farmers to invest in the equipment. Tractors reduced the need for hired labor, were more versatile than horses or mules, did not consume fuel unless they were working, did not require veterinary treatment, and never dropped dead of unknown afflictions or from overeating. The mechanization of cotton harvesting and cultivating accelerated crop diversification already taking place in the South Plains area by making cotton a more attractive cash crop. Partly as a result, cotton began replacing sorghum and wheat at the same time that new developments appeared in irrigation. Although of only local interest at the time, groundwater irrigation on the South Plains came to play a major role in the future of the Canadian River.[9]

The sandy soil of the South Plains, combined with hot, dry late summer weather and new strains of short-staple cotton adapted to the Llano Estacado, encouraged the rapid spread of cotton into the region. Prices for the fiber remained steady through the decade, providing a better return in some cases than did small grains. In 1921 short-staple cotton of fair to middling fair quality (the best) brought 9.2¢ per pound. Over the decade the price increased to range from 15.9¢ to a peak of 32.3¢ per pound, while cottonseed, used in industrial oils and animal fodder, brought over

$25 per ton. Cotton gins opened in Plainview, Lubbock, and other towns on the South Plains, creating new jobs. Cotton provided the best yields when irrigated, and the "shallow-water belt" of the Llano happened to coincide with the best cotton land, a happy development for farmers and well drillers alike.[10]

Irrigation from groundwater was not entirely new to the High Plains, but technological developments made it much more feasible after the 1910s. The centrifugal pump and so-called pitless pump were more durable, required less maintenance, and used less fuel to run per horsepower than did other models. They also had greater lifting capacity than earlier pumps, making deeper and higher-output wells possible. However, the technology spread slowly at first. Wells, pipe, motors, spare parts, and fuel cost money that banks did not always want to loan. Developers also had to teach farmers the best ways to use irrigation water, notably showing them to not wait until crops started dying before irrigating the fields, much as had happened at the start of surface water irrigation in the 1800s. Despite irrigation proponents' enthusiasm, some groups opposed irrigation. As agricultural prices declined in the early part of the decade, Department of Agriculture extension agents began actively discouraging client farmers from irrigating or from expanding already irrigated acreage. Further limiting the prospects for a well-watered future, not all land was suitable for irrigation: it had to be very flat but well drained, with fairly "shallow" (easily accessible) water of sufficient supply to make pumping worthwhile. Even so, the area between Hereford and Lubbock, especially around Plainview, Lubbock, and Shallowater (Deaf Smith, Lamb, Bailey, Hale, Parmer, and Lubbock Counties) seemed particularly well suited to cotton and irrigation both, and by 1929 thousands of acres of fleecy white bolls were picked and sent to the gins for cleaning and shipping. The decade was (relatively) damp, and land salesmen touted the region's freedom from the boll weevil, the availability of inexpensive land, and the steady prices for cotton. While agriculture contracted in some areas in the mid-1920s, things seemed to improve for the High Plains. [11]

Wheat prices increased in the closing years of the decade, raising eyebrows and concerns both in the Panhandle and in Washington, D.C. Better seeds, increasing use of technology, and a series of "wet" years on the High Plains encouraged farms to expand, especially monocrop acreage that focused on cotton and wheat. As historian Donald Worster noted, mechanization lowered costs of production enough so that area farmers could make a living on $1.03 per bushel because they needed only $.40 to cover the costs of fuel, fertilizer, and equipment and labor. The price supports put in place in 1929 also encouraged speculation in land and grain, leading some people to grow wheat for the sole purpose of raising the price of the land so they could sell it for a quick profit. A farmer from north of the Canadian River, Lucien Burnett, recalled that most of the rangeland around his place yielded to the plow only in 1929 as people planted in anticipation of a new railroad line predicted to arrive in 1930.[12]

Not everyone was comfortable with the spreading sea of wheat. Worried by the wheat monoculture, H. M. Dainer wrote a letter to the *Amarillo Daily News* urging a return to the safety-first practice of stock farming, complete with "the cow, the sow and the hen." The rise of suitcase farmers exemplified the growing dependence on wheat. The increasing amount of suitcase farmers troubled long-time residents. A number of these suitcase farmers were pure speculators, breaking the ground, growing wheat, and then selling the crop on the market just so they could offer the now more valuable land to another wheat-chaser. Some resident farmers watching this potentially unhappy trend felt that a person should live on and observe the ground that he or she worked, keeping an eye on weeds, erosion, and fences. Absentee owners and speculators did not contribute to settling the area, made no long-term commitments, and drove crop prices down for those who were trying to live on the land and raise families.[13]

Many state and federal agricultural specialists assisted the ongoing westward march of winter wheat (and land speculation) by promoting water-saving farming techniques. One technique involved using a one-way plow, so-called because the ground-cutting disks were all angled in the same direction, in order to till the soil so that no moisture-stealing weeds grew when the ground lay fallow between crops. The new tractors pulled this wonder plow over thousands of fresh acres as wheat prices rose a little. Although encouraged by some agricultural experts, this rapid expansion of wheat into marginal areas of the Texas Panhandle, eastern New Mexico, eastern Colorado, western Kansas, and into the Dakotas and Montana, termed the "great plow up," worried some of the old Progressive conservationists as much as suitcase farming worried settled farmers. The conservationists watched the economic and social woes developing in agriculture over the course of the 1920s and wondered if rather than being purely farming problems, these were natural-resource problems akin to strip-mining or to clear-cutting timber.[14]

Concerned conservationists asked: did farming make the best use of the soil and water of the South Plains and High Plains? Hugh Hammond Bennett, a soil scientist who began his work in the Deep South, thought

not. Bennett, along with a growing number of observers and researchers, concluded that the agricultural lands of the United States needed
to be studied and classified by experts who would then decide the use of
those lands. According to the experts, any ground that did not produce
enough to balance the effort and resources put into it was "economically
marginal" at best. Farmers should stop growing grain and instead use the
land for timber, pasture, or some other crop that would pay. Bennett was
not alone in this. As early as 1924, the *Yearbook of Agriculture* from the
USDA included a long and detailed article outlining how farming and
industry should both be rationalized, a plan that included pulling the
plows back to the east and letting grass and grazers return to much of
the High Plains. But Bennett's concept of tying the "farm problem" to
the rational scientific use of natural resources was very progressive and
very new. As a result it received at best mixed responses from the general
public and from farmers, especially during the latter part of the 1920s.
Farmers and ranchers on the High Plains wanted to receive advice on
keeping prices up and pests down, not to be told that they did not know
what they were doing and that experts from "back east" knew the land
better. Other people, including a governor of New York named Franklin
Delano Roosevelt and a professor of agriculture at Columbia University
named Rexford Tugwell, listened to Bennett and took notes for future
use. Meanwhile, another valuable natural resource had been discovered in
the Panhandle; one that, like wheat, produced mixed results for the land
and the landowners.[15]

Quick money and equally rapid bankruptcy accompanied the oil boom
into the Texas Panhandle. As early as 1905 the geologist Charles N. Gould
had noted some potentially oil-producing rock formations as he surveyed
for groundwater. However, no serious exploration took place because of
the cost and a lack of markets for whatever oil might be found. That
changed in the 1916, when M. C. Nobles hired Gould to come back and
remap the area, this time looking for oil instead of water. In October of
that year, a group of businessmen and ranchers hired the Hapgood Drilling Company to drill test wells on Lee Bivins's LX Ranch and Robert
Masterson's JY property, both within the Canadian Breaks. After drilling
a number of dry holes, Hapgood's roughnecks brought in the Masterson
Number One in December 1918. The well produced "fifteen million cubic
feet of natural gas per day" from a depth of 2,605 feet below the surface.
Natural gas was not "black gold" but it was a good sign, and a market had
already developed for natural gas as heating fuel. The gas strike encouraged the Amarillo Oil Company to fund more wells.[16]

Exploration produced mixed results at first. After striking gas several more times, the Gulf Production Company hit oil in Carson County, east of Amarillo, on May 2, 1921. It was a small pool at a depth of 3,052 feet (930 m) below the surface of the ground (roughly 400 feet [122 m] above sea level), and the well produced only 200 barrels (8,400 gal; 31,800 l) per day. Gulf explored a little more and brought in a gusher just north of the Canadian Valley in Hutchinson County not long after. However, the geology seemed unfavorable because the wells kept running into granite, and major companies such as Gulf and the Texas Company turned their attention elsewhere in Texas, Louisiana, and Oklahoma. As later studies would disclose, the granite formed the buried Amarillo mountain range. The range's rocks trapped oil on its northwestern and eastern flanks and marked the northwest edge of the vast Anadarko Basin of Oklahoma and northern Texas. After the majors left, independent drillers, called "wildcatters," continued exploring the Canadian Valley and surrounding area and eventually hit what became the Dixon Creek segment of the Panhandle field in 1925 in southeastern Hutchinson County. But no matter who did it or what they found, all this drilling required water.[17]

No one could drill for oil without having a reliable supply of water. The cable tool rigs favored by wildcatters for their low cost, their ease of operation, and the quality of the chips they produced could not function without having muddy water in the hole. The drill bit worked by pounding up and down, breaking the rock in the path of the well. Water served to lubricate, to loosen the rock chips, and to flush the chips out of the hole. Drillers' manuals recommended having no less than one barrel (42 gallons) of water in the hole at all times. Since drillers bailed the water-rock-mud mixture out of the hole after the drill dug eight or nine feet, making a 2,000-foot-deep well required at least 9,300 gallons of water, plus more for firefighting, rinsing off rock chips, and sanitation. No water meant no progress, and area newspapers noted several occasions when drought or freezing weather brought work to a halt. In one instance drilling stopped on the Tuck-Trigg well in Carson County for two weeks in April 1920 because the nearby creek that supplied the drilling site had run dry.[18]

All that drilling consumed a considerable amount of water. By the end of 1925 more than eight hundred wells dotted the Panhandle skyline, meaning that if each well had stopped at 2,500 feet below the surface, more than 9.3 million gallons of water had been poured into the dry Panhandle ground. For comparison, residents of the largest city in the area, Amarillo, used roughly a million gallons of water in one year. Most people did not seem to mind the soaring water consumption, especially the new ones who flocked to the area as boomtowns appeared, catering to every need (and most desires) of the men working in the new oilfields.

In 1925 Borger, Texas, did not exist. Instead, ranchland covered the south side of the Canadian Breaks, not far from where the river bent from northeast back toward the east. An oil strike on the Dial Ranch on the north side of the river in Hutchinson County resulted in a well that produced 135 barrels of oil per day, and on January 11, 1926, the "Borger rush" started as drillers hurried to the area. Asa Phillip "Ace" Borger came from Missouri with the oil business but decided that there were surer ways to make money. He paid six thousand dollars for part of a ranch, then marked out a 240-acre (97 ha) town site. He started selling lots in March 1926 and, according to tradition, made one hundred thousand dollars by the time he closed the door the first day. Tents and shacks sprouted, and soon more than ten thousand souls walked and drove along the dirt (and mud) streets. Borger's reputation grew as fast as the town, and law enforcement failed to keep up with the chaos, leading Governor Dan Moody first to send in Texas Rangers and then to declare martial law in 1929. The second time he sent in rangers, backed by the National Guard, was to reestablish order following the assassination of the district attorney "by person or persons unknown." Ace Borger himself became a victim of the violence, shot dead in 1934 by county treasurer Arthur Huey. Amarillo's residents must have looked at the new town and shaken their heads, wondering if the Wild West had returned. Newcomers looked at the town and wondered if it had already gone to perdition—literally.[19]

Anyone trying to "clean up the town" in the late 1920s would have taken one look at the billowing black clouds drifting over Borger and fled. The natural gas associated with the Dixon Creek field and others nearby was termed "sour gas" because of the smell of the chemicals in it. Since it could not be used for heating without pretreatment, companies produced carbon black from the "raw" gas instead. Carbon-black manufacturers made their product by burning the gas in a low-oxygen atmosphere, then scraping out the residue. Other manufacturers bought the black powder to use to reinforce natural rubber products, including tires and belts, and to make black inks and paints. A fair amount of the powdery material escaped the plants, and photographs of the town taken between 1926 and the early 1950s often show low black clouds blowing away from the carbon-black factories. Women living around Borger coordinated doing their laundry with the wind direction so that a minimum of residue would fall onto the clothes drying on the line.[20]

Figure 2: A little too close to water? Wells near Borger in 1926. Used with permission of Nita Stewart Haley Memorial Library, Midland, Tex. Image JEH-J-5.7.

Producers burned off—flared off—the waste gas and chemicals at the wells and refineries, adding to the industrial atmosphere. When Edna Quisenberry Jackson moved into Borger as a child in 1939, the sight terrified her. "We thought the town was burning. Carbon black! There was a deep snow with carbon all over it. The flares and the raw gas burning made a stench worse than the pigpens on our farm." She and her sister thought her mother and stepfather had brought them to hell. Between carbon black, flared-off gas, and the residue of gushers and other oil wells, the air, soil, and water pollution around the Panhandle fields was notable but typical for the time. The residents did not especially care for the muck, stench, and soot, but considered them the price of prosperity. And not everything suffered. Elanora Engle Walker remembered gathering large amounts of wild plums from the Breaks just north of town in 1928 and 1929. And the residents of Borger had another more pressing problem than just pollution, a problem that they shared with Amarillo: the problem of water.[21]

Borger residents needed water. The cost of pumping the muddy, often brackish river water up over 400 feet (122 m) to the town site, then treat-

Figure 3: Carbon-black plant near Borger [1950s?]. Used with permission of the Hutchinson County Historical Museum, Borger, Tex.

ing it, eliminated the Canadian as a source of drinking water. Amarillo's practice of pumping its treated sewage into the Canadian upstream from Borger, along with the numerous oil and gas wells located on the banks of the river and in the riverbed, provided still more reasons not to drink from the stream. Oil gushers, although not common, wasted petroleum and soaked the soil with heavy oil, some of which probably washed into a few of the streams as well as into the Canadian. As a result, Borger residents tapped springs and clean tributary streams via the services of a water wagon that came by once a week on Fridays. According to early Borger resident Herbert C. Wilson, the waterman charged a dollar for one barrel, or you could buy it by the gallon from some merchants in town. J. D. Williams sold water for ten cents a glass at his store. Water cost more than the whisky (another product of the Canadian River Breaks) that flowed rather freely despite Prohibition, and inspired almost as many efforts to obtain.[22]

Phillips Petroleum, as desperate for water as the rest of Borger, hired self-taught water engineer Paul "Silent" Endicott to see what he could find. Endicott located a good water-bearing formation near Dixon Creek and drilled several productive wells. Phillips took what it needed and sold its excess water to town residents at a reasonable rate. This worked well enough, although as time passed some Borger leaders worried about being dependent on the town's largest employer for their drinking water. Other refineries and plants, most notably the carbon-black plants, sprang up to make use of the oil and superabundant natural gas from the Panhandle field, and all these industries needed water. It became quite apparent that eventually the water wells might start running dry. Farther to the north and east, the town of Pampa also grew with petroleum-related businesses and plants. As in Borger, Pampa's town leaders began looking at other possible water sources. Pampa had always depended on fairly deep wells for its water, making extra supplies rather expensive to obtain. Amarillo, perched on the watershed divide between the Canadian and Red, ran short of water as well, even as it boomed along with Borger and Pampa.[23]

Amarillo had faced water shortages interspersed with floods ever since its founding. The 1920s were no exception, and as the city grew, more and more people bought amenities such as flush toilets and washing machines. Industries related to oil, cattle, wheat, and manufacturing developed in an area on the north and east of downtown called "industrial city." Nine thousand nine hundred fifty-seven people called Amarillo home in 1910, while 15,494 responded to the census enumerator in 1920 and an amazing 43,132 resided within the city limits in 1930. In ten years more than twice the number of people needed to drink, wash, and flush, creating a major supply and disposal infrastructure problem for both city manager Jeff Bartlett and the three-member city commission to solve. By 1925 the private company that had been providing municipal water service could not keep up with demand, leading to numerous complaints about low pressure at the taps and rising fire-insurance key rates for businesses in the city. At the same time, Amarillo negotiated to purchase water from under the Word Palo Duro Ranch in southern Randall County. Now, all that the city needed to do was to get the well water 15 or 20 miles uphill into town and to find a way to dispose of the "gently used" water.[24]

Solving the inflow and outgo problems took the better part of four years, from 1925 to 1930. City Power and Water, a local branch of water and power company Southwestern Public Service, in July 1925 offered to sell the city its distribution lines for $750,000 or to add the central pumping plant for a total of $2.5 million. This was more than Amarillo's leaders (and many taxpayers) were willing to pay for the system. Negotiations over the price and worth of the infrastructure, water rights, and wells continued for several years but ultimately failed, leading the city of Amarillo to condemn the plant and pipes, eventually agreeing in March 1927 to purchase everything. At the same time, the city drilled twelve wells in the Randall County field, pumping the water into a new municipal reservoir lake and thence into town. The lake became a popular recreation area, complete with fish, shade trees, and other amenities. The piscatorial population eventually succumbed to one of the recurring droughts, fished out with the utilities' manager's blessing lest they die and cause a health hazard for picnickers.[25]

Disposing of drought-stranded fish came easily compared to disposing of urban sewage. The city banned outhouses in the early 1920s. Because most of the city still lay along or just north of the Canadian-Red River divide, once everyone's pipes connected properly to the system, gravity-flow disposal of storm water and sewage was fairly simple. The problem lay with the next step. The plans for city development included draining the raw sewage and rainwater into East Amarillo Creek and thence into the Canadian, where dilution took care of the effluent. This was a common practice and in line with the saying that "dilution is the solution for pollution." The people living beside the creek in the River Road community respectfully disagreed, pointing out that storm-water runoff already eroded their property and property values even before Amarillo added its effluent to the mix. When asked about the matter, the Texas State Health Department suggested not dumping city sewage into Amarillo Creek or the Canadian River, in part because the river's waters belonged to the state. After some debate, in late 1927 Mayor Lee Bivins granted the city several acres on his ranch for a waste treatment system similar to that in Lubbock. Lubbock, to the south, used a sewage farm to take care of its waste, and experts in the area considered that city's system to be a model of efficiency and sanitation. Amarillo's waste went into Imhoff tanks, very large versions of a septic tank, and the city used the results to irrigate pasture for horses and cattle as well as for growing alfalfa. This served for the moment, although complaints about low water pressure continued to plague the city. It was obvious that wells in Randall County would not be enough for the future, despite optimistic declarations made by members of the Board of Water Development on January 3 and February 16, 1925.[26]

When Mayor Lee Bivins took office in 1925, among his first acts he ordered the city manager, Jeff Bartlett, to look into obtaining water rights on the Canadian in both Texas and New Mexico. As owner of the LX ranch along the Canadian, Bivins knew very well how much water could be had if only it were captured and held for later use.[27]

In this Bivins echoed a better-known contemporary proponent of both river capture and railroads, Albert Sidney (also known as A.S. or "Sid") Stinnett. Stinnett had moved to the area from Fort Worth in 1905 to sell cottonseed cattle feed. In the course of his dealings with area ranchers and cottonseed and cottonseed oil shippers, Stinnett noted how the lack of transportation hindered growth, especially north of the Canadian River, and how the lack of surface water supplies made things even more difficult. Much of the area along and north of the Canadian was in a "deep water" region where one had to drill a deep (and thus expensive) well to reach dependable water from the Ogallala Aquifer's sand and gravel. Land values remained too low to justify drilling the expensive irrigation wells because there was no way to get crops to market from north of the Canadian, unless one lived on the old XIT lands or near the Canadian River Bridge built north of Amarillo in 1912. To Stinnett, the waters of the Canadian, including those that rolled downstream in the flood of 1914, were wasted and should be conserved for irrigation and industrial use while railroads tied the northern side of the stream to the rest of the world.[28]

Stinnett was one of the first to seriously propose tapping the Canadian as part of the larger development of the Panhandle. He initially broached the idea in 1918 as a presentation to the Amarillo Board of City Development. His ideas struck a chord, and the next year Amarillo business leaders elected Stinnett president of the board. He also helped found the Panhandle Plains Chamber of Commerce that same year. Shortly thereafter Stinnett became the head of the Industrial Commission of the City of Amarillo, lobbying railroads for another north-south line from Amarillo. While working with regional leaders, he also served at the federal level as a member of the Arkansas River Commission. In 1923 Stinnett introduced the idea of harnessing the Canadian for flood control and local development as part of a larger national Arkansas Basin river development plan. The well-known civil engineer and irrigation surveyor Vincent Jones of Las Vegas, New Mexico, son of U.S. senator Andrieus A. Jones, suggested that the idea was feasible.[29]

After some consideration, the Chamber of Commerce and others hired consulting civil engineers to look at what would be involved in damming the Canadian River. The engineers delivered their report in January 1924

Figure 4. Two-horse power: The Borger Car Ferry in the 1920s was not what Stinnett had in mind as a reliable river crossing. Used with permission of the Hutchinson County Museum, Borger, Tex.

and proposed a dam just west of the Texas–New Mexico border, probably near what would be later called the Dunes Site. Canals would carry the stored floodwaters along the southern edge of the Breaks, providing much-needed irrigation waters as far to the east as Carson County. The engineers did not have enough data to be able to guarantee how much water would get that far or what the final cost would be.[30]

Nothing more came of the study. Although the documents do not say exactly why, the reasons are easy to imagine. The cost and scale of the proposed project were almost unimaginably large, larger than the Panhandle could support even in the first flush of the oil boom. Enough groundwater existed to meet everyone's needs, even if it was relatively expensive for a prosperous city like Amarillo, let alone for smaller towns. Dry-land wheat brought in plenty of cash most years, making irrigation an unnecessary expense for farmers. Those owning the land along the river proper, the Matador Land and Cattle Company, Julian Bivins, Lee Bivins, Robert "Bob" Masterson, and others, had little interest in losing their best-watered lands to a reservoir no one wanted that badly, especially one that would endanger their oil royalties. The dam and canal

RED WATER, BLACK GOLD

system remained an idea to be investigated for the future, when money and need and imagination might combine to make taming the Canadian feasible.

Meanwhile, A. S. Stinnett put his money where his mouth was. He lobbied for a railroad crossing for the Canadian east of the one near the old settlement of Tascosa, eventually paying for part of the construction himself in order to connect Hutchinson County and Amarillo with Liberal, Kansas, in 1925. The arrival of the railroad sped the settlement of Hutchinson County as land prices rose, leading ranchers to break up their properties and sell to the newly arriving wheat farmers. Stinnett, along with Ace Borger and J. T. Peyton, founded the town of Stinnett on August 16, 1926. The town grew quickly. It served as a rail depot on the Chicago, Rock Island and Gulf Railroad before prospering as an oil boomtown and deposing Plemons to become the county seat. Two years later, the world's second-largest carbon-black plant (at the time) opened for business in Stinnett. Agricultural and industrial development along the new rail spur boded well for the area, provided that water could be found enough to keep towns and crops growing and wildcatters drilling.[31]

According to Thelma Meredith Lofgren, A. S. Stinnett and the Amarilloans were not the only people thinking about a dam on the Canadian. Her father, Alson Asa "Double A" Meredith, worked for Gulf Oil near Borger in the late 1920s. In an interview with Terri Maxey in 1973, Mrs. Lofgren explained that "[m]y earliest recollection of even a comment of a dam across the Canadian River dates back to the Model T era. There was a group of men that did a lot of hunting . . . Bob Ames, George Maggard, Jim Maggard and Dwight Newby, these were just a number of businessmen, particularly Rotarians, in Amarillo and I remember going out to the Canadian River in a Model T with dad and hearing them talk of a dam across the river to hold the water. Because a number of times when we would have rain we would go out there and there was water at the bottom of the bridge just rolling." Meredith filed his thoughts and ideas away when Gulf Oil transferred him to Plainview, Texas, seventy miles or so from the river. Meanwhile, life moved along in the Breaks.[32]

To the west of Amarillo, visitors to the Canadian Valley would have been hard-pressed to tell that they were still in the twentieth century. Although farming expanded onto the uplands in the northwestern Panhandle, the Canadian Valley in Oldham and Potter Counties remained ranch land. The Matador Ranch's Alamositas Division operated much it had when it was part of XIT, with the addition of a few cars and a pickup truck. Hereford cattle and ranch horses grazed the rough pastures along both sides of the river, and the seasonal round of spring roundup and branding, shipping cattle to market or to northern pastures in South Dakota and Canada, the fall roundup, winter feeding when necessary, and riding bog in the spring along the river's banks continued.

James Stevens, the division manager, noted dry weather and wet, floods and droughts, in his daily log. He also recorded multi-week visits by the Matador's Scottish manager, Murdo McKenzie, and guests who came to hunt the ducks, quail, and occasional deer that still resided in the Breaks. And Stevens and his assistants noted when the autos had to be pulled across the Canadian by pickup or by horses, or had to be dug out of the mud along Rita Blanca Creek, or drive from Channing north of the river to Vega on the south via Amarillo because of high water. The ranch grew some of its own food, raised pigs that were slaughtered once cold weather set in, and looked as much to Dundee, Scotland, as to Amarillo or Vega for news and directions. So long as cows were cattle and the Canadian was unruly, life on the Matador, on the LX, at Boys Ranch, and on the other ranches downstream of Amarillo would continue much as it always had. But changes were coming to the river itself, instigated by "equinoctial floods" earlier in the decade.[33]

In 1923 the Canadian River boomed and roared, starting a chain of events that would lead to the first Canadian River Compact between New Mexico, Texas, and Oklahoma, the first dam on the western part of the river, and the re-ignition of interest in taming the Texas reach of the stream. A flood, a compact, a drought, and dust storms, along with the ambitions and visions of Panhandle residents and national conservation experts, moved the next chapter of the Canadian River story as people fought to save water in order to save the soil and their way of life when severe drought returned to the Canadian watershed. The 1890s had been dry, but the 1930s would be worse.

Chapter 2

A SMALL RIVER IN THE BIG LEGAL PICTURE

O n October 14 and 15, 1923, as conservation-minded Texans and New Mexicans considered the best ways to use the Canadian River's waters, the Canadian and its major tributary devoured bridges and bottomland in Oklahoma, turning the town of Woodward into an island and sending refugees streaming up to the high ground around the state capitol's hill in Oklahoma City. When the sun rose on October 16, its light revealed damage that extended across the entire width of the state. For a week, no one could cross the state north-to-south, no trains ran, and damage done to Oklahoma City's municipal reservoir endangered the city's water system. Oklahoma's leaders decided that the time had come to do something about the Canadian's rampages. The nation was taming other streams and now it was the Canadian's turn, if Oklahoma had any say in the matter.[1]

Oklahoma's need for flood protection and Texas's and New Mexico's desires for drinking and irrigation water led to the first Canadian River Compact and spurred more investigations into damming the stream. Taming the river led in turn to considerations about the legal status of the Canadian's waters. At the same time, tristate efforts to conserve the Canadian's waters fit into a national trend as basin-wide flood and channel protection became a national policy. During this time states worked together to create a number of river compacts, the interstate treaties dividing up the waters of the Colorado, Pecos, Rio Grande, and other western streams. The Canadian River Compact of 1928 serves as an example of how regional interests within the basin shaped the final treaty and the problems of building a large dam on a (usually) small stream.

The flood of 1923 was what residents of Tascosa would have called an "equinoctial flood," one that came around the time of the autumnal equinox. The 1904 flood came at a similar time and the two resembled each other in scale, but not in damage. The river caused far worse damage in the New Mexican reaches of the river in 1904, killing at least seven people, washing away small towns, and ripping out bridges upstream of Texas. This time, although the water rose in New Mexico and Texas, Oklahoma suffered the brunt of the flood damage. A series of storms brought steady, heavy rain to eastern New Mexico and the Texas Panhandle, rain that filled both the North Canadian (or Beaver River) in far western Oklahoma and the main stream of the Canadian in Texas. High water turned Woodward, Oklahoma, into an island in the North Canadian for a day and a half. The South Canadian destroyed valuable bottomland farms, ripped away railroad and wagon bridges, and almost turned Norman, south of Oklahoma City, into an island again. The state estimated that the damage would total $75 million in crop losses, ruined farms, and infrastructure damage.[2]

Once again Oklahoma had more of Texas's and New Mexico's water than it wanted, and the time had come to fix this problem. However, it was not just the Canadian's branches that caused havoc and heartache in the state: the Cimarron and Arkansas needed domestication as well. Governor Martin E. Trapp contacted people in Texas and New Mexico, and also Nebraska, Kansas, and Colorado upstream, and in Arkansas downstream. The Oklahoma governor wanted to create a body to organize flood-control projects in the entire Arkansas River basin, including the Canadian, Cimarron, and other tributary streams. After some discussion and following strong lobbying by the Tulsa Chamber of Commerce and other organizations, in 1924 the states formed the Arkansas River Basin Association with the goal of creating river compacts and promoting flood control on all tributaries of the Arkansas.[3]

Conflicting state water laws and the realities of life west of the 100th meridian led to the creation of numerous river compacts. Eastern states often reached agreements about pollution control, but found no need to divide up or apportion the waters of, for example, the Ohio or Shenandoah, because there was plenty of water and because of the nature of water laws in the eastern states. The common law was *riparian* law, which regulated usufruct of the river. Individuals and organizations owning land along the stream had the right to use the water (usufruct) so long as they did not injure the rights of people downstream by diverting too much water or by polluting the river. River laws dealing with dams and weirs focused on right-of-way and flood questions, not on how much water each property owner should have, and made no distinction between those

Figure 5. Not a good day to wade: A flood in 1923. Used with permission of the Panhandle-Plains Historical Museum, Canyon, Tex.

who had just bought land and those who had been using the river for several decades. Discussions of river water use did not consider irrigation by landholders without riverside land, in large part because water rights depended on ownership of riverbank land. Land rights and riparian water rights remained connected.[4]

In contrast, western states favored the law of *prior appropriation*, a doctrine derived from California's mining districts and sometimes attributed to earlier Spanish water laws. Also termed "first in time, first in right," prior appropriation doctrine holds that the first user to appropriate water from a stream and put the water to "beneficial use," however defined by the state, owns the first or most senior right to that water, so long as he or she continues to make beneficial use of the full amount claimed. The next person to put water to use has a junior right, even if they are upstream of the senior, and in times of shortage the junior cannot use their water right until the rights of the senior have been fulfilled. People considered irrigation to be a very beneficial use, and rights for irrigation dominated the list of water claims, followed by industrial uses such as iron smelting. Unlike riparian-right holders, someone with an appropriation right could move his water away from the streambed and riverbanks, a feature found both in mining and in Spanish water laws. Colorado, New Mexico, Wyoming, Arizona, California, and other western states included prior appropriation as the official state law, and the state engineer registered those rights. The engineer or a state court determined seniority in times of conflict. The state engineer also decided just how much water was in the streams to be apportioned, effectively setting a cap on the number of water rights available.[5]

Texas and Kansas initially used riparian law, but over time both shifted to a combination of riparian in the wetter parts of the states and prior appropriation for the more arid regions, such as the Canadian and Pecos basins. Texas, in its post-Reconstruction constitution and more explicitly in chapter 88 of the 1889 General Land Laws, recognized both riparian and prior appropriation rights depending on what part of the state (humid or arid) was in question. Texas also claimed all previously unclaimed navigable surface waters and their associated watercourses. Further revisions in 1895 reconfirmed the state's power to regulate apportionment of surface water, which the state "held . . . in trust for the whole people." That explains why Amarillo mayor Lee Bivins inquired with his city manager about obtaining water rights to the Canadian in 1925, as a way to preempt later claimants should it become possible to use the river's waters. Only one claim on the stream in Texas existed at that time, for irrigating a pasture downstream of Amarillo near what would become Borger. As a result, surface water rights within the state were not a problem for would-be Panhandle water conservators.[6]

Although not yet a consideration in the Canadian River watershed during the 1920s through the 1940s, groundwater rights in Texas deserve some discussion. Neither Texas's water-regulating agency, the Board of Water Engineers, nor state statutes regulated groundwater. In 1904 the state supreme court ruled in *Houston and T. C. Railroad vs. East* (98 TX 146) that the ways of groundwater were too mysterious to be regulated by the court and that a property owner had full right to all uses of all groundwater on his property. As a result the Board of Water Engineers, created as part of the major reworking of state water laws in 1913, had no oversight of groundwater aside from springs and those waters found under the bed of a stream that were obviously normally part of the stream. In contrast, New Mexico assigned groundwater to the supervision of the state engineer, who allotted it much as his office allotted surface water, including restricting pumping so as to limit interference with neighboring wells. In the 1920s through the 1940s, this peculiarity of Texas water law remained more or less irrelevant for Panhandle and South Plains residents. Only later would use and overuse of groundwater affect the Canadian River directly by reducing the flow of springs and indirectly by forcing cities well outside the watershed to turn to the Canadian for municipal water supplies. It is worth a further small diversion to compare groundwater law with the development of the regulations concerning the state of Texas's other law-of-capture resource: oil.[7]

The economic value of petroleum and natural gas led to regulations and state-mandated conservation. From the time of the first oil boom in Texas in 1903 until the mid-1930s the law of capture ruled oil fields just as it ruled water wells. Oil in the ground belonged to whoever owned the mineral rights and pumped it first, leading to the well-known drilling frenzies of the 1910s and 1920s. It also meant that the discovery of a new field sent oil and later gas prices plummeting as too much crude and refined petroleum flooded the markets. Unable to store gas in the ground without losing their rights to it, oil companies burned off phenomenal amounts of natural gas, in the process decreasing the pressure within the oil-bearing formations (and shortening the life of the field) and wasting otherwise useful fuels. Through a series of trial-and-error attempts at regulation, the Texas Railroad Commission and the state legislature eventually found an acceptable method of resource conservation called *proration*. Put into place in the 1930s to shore up the Depression-wracked oil industry, proration in effect rationed oil by restricting the number of oil and gas wells per acre of ground and by limiting the amount each well owner could pump from a given field in a year. No longer were people punished for conserving oil and gas in the ground. This led to better management of the resources and encouraged wiser, more careful development and use of subsequent oil fields and gas finds. But the precedent set by oil-field necessity never reached the groundwater courts, except in a very limited local-option version through the creation of underground water conservation districts. That groundwater conservation development remained far in the future for the High Plains of the 1920s.[8]

During the Roaring Twenties and Dirty Thirties, surface-water rights occupied plains residents' attention much more than did groundwater rights. The great concern for water developers and dam proponents remained prior claims to the Canadian, and no doubt the Texas and New Mexico boosters felt relieved when they each found no private claims to the Canadian's waters in their respective states. Each state's rights to allocate the river's flow appeared clear and unencumbered, within the limits of what the environment provided in precipitation. But rights among the three interested states were another matter entirely.

River compacts in the United States had a relatively uncomplicated history prior to 1925, as far as matters of interstate and water-related laws go. According to water lawyer and historian George W. Sherk, all river compacts and all subsequent litigation over those compacts in the United States derived from three causes: too little water moving from one state to another, polluted waters, or too much water traveling downstream. Although the federal government retained control over all "navigable waters," the federal government had no say over how states divided shared waters or how the states regulated chemicals and dirt flowing downstream. Congress initially chose not to legislate interstate water apportionment. After 1900 the states, especially those west of the 100th meridian, jealously guarded the privilege of negotiating retention and release of stream waters. Only in cases where international conflicts arose from a state's proposed water development did the federal government involve itself as a party to water compacts, for example the compact between Texas, New Mexico, and the country of Mexico that regulated the flow of the Rio Grande.[9]

Representatives completed the first Canadian River Compact, a treaty negotiated between the states of New Mexico, Texas, and Oklahoma that divided the waters between the states, on December 31, 1926. The Canadian River Compact contained features similar to some in the more famous Colorado River Compact. As article I, paragraph (a), of the Canadian River Compact explained,

> The major purpose of this compact is to provide for the control of the flood waters of the Canadian River . . . and the equitable division and apportionment of the use of the water of said River System impounded hereunder, so far as may be, and the *equitable apportionment of the cost of control of said river, according to the benefits received from the control of said river* and the use of the waters thereof. [emphasis added]

Article II defined terms used in the compact. The next section stated that New Mexico and Texas would create a commission to determine how the upstream waters would be divided between them, then went on to describe the limitations placed on such impoundments and releases of the Canadian's waters. It is important to note that in no place in the compact of 1926 did negotiators insert any description of how much water each state could keep—no acre-foot limit or minimum—except to say that the states should not hold so much water back that it forced them to release extra during a flood and so make flooding worse. The compact also stated that in no case would downstream states request more water than could "be reasonably applied to domestic and agricultural uses in the lower state." In this the compact was very much like most other river agreements except for the lack of specific water amounts.[10]

The provision for dams and dam-construction funding made the Canadian River Compact unique. Article I described how the costs would be apportioned among the three signatory states. Article V specified that dams would be built "at sites to be selected on the main stem of the Canadian River, and on Ute Creek and on Pajarita [*sic*] Creek," in New Mexico. The combined dams could hold up to 850,000 acre-feet of water for irrigation, and "such additional capacity for flood restraint as may by the Commissions of the lower states be deemed adequate to restrain and control the flood flow of said stream system." Translated into English, should New Mexico build a flood control dam, Oklahoma, and Texas would each pay roughly one-third of the cost, depending on the financial benefit from flood protection in each state. The same applied to water-retention projects in Texas, and again the compact required no minimum flow-through. In this combined-payment plan the three states looked back to geographer and conservationist John Wesley Powell's ideas for political and hydrological development by watershed and not by state. Residents of the Canadian watershed as a whole would pay for flood control projects in the basin as a whole, rather than only for projects within their own states. Specific water-users and the state water districts were to pay for any irrigation or domestic diversions from the created reservoirs, that is, the pumps and canals, but the dams were different. The stream caused problems for everyone and so all would help control it, or so people hoped in 1926.[11]

As with any treaty, the governors' representatives submitted the Canadian River Compact to the legislative and executive branches of the states involved before sending the document to the U.S. Senate for final approval. New Mexico and Oklahoma ratified the compact quickly. The Texas legislature took a little longer, sending the treaty to the governor on March 9, 1929. However, Governor Dan Moody vetoed the measure. Officially, he had concerns about the costs to the state and about the lack of study done concerning benefits for the state, and he questioned the authority of those who would administer the compact.[12]

Although disappointed, Oklahoma and New Mexico acted as if the compact had been passed and approved and began making dam plans. Residents of the Panhandle also ignored the governor's decision. Texas's compact negotiator, A. S. Stinnett, attended the Canadian River Commission meetings as a nonvoting observer, bringing back news of developments. The New Mexico state engineer, Herbert Yeo, began serious studies and preliminary plans for a series of dams on the Canadian and its major tributaries, Pajarito Creek and Ute Creek. At the states' request, the U.S. Geologic Survey installed river gauges on various parts of the Texas and New Mexico stream in order to record the flow so that engineers could better determine baseline parameters for dam design. Likely dam locations on the main channel included the confluence of the Canadian and Conchas Rivers, somewhere in the narrow valley near Ute Creek's junction with the Canadian, near Tascosa, Amarillo, and at the bend near the new town of Borger. Which led to the question: who would build the dam(s)?[13]

Neither Texas nor New Mexico, nor Oklahoma for that matter, was a "rich" state at the time. Texas was the best off of the three because of its land and royalty income, although Oklahoma also prospered somewhat from oil revenue. Minerals found on state land or in state riverbeds belonged to the state, and the legislature put the funds aside for education as well as using some of the income. New Mexico and Oklahoma possessed less state land and could only watch as oil companies developed finds on private, federal, and American Indian properties. New Mexico lacked the other states' large petroleum incomes, although exploration in the southeastern part of the state suggested that an oil strike might be in New Mexico's future. For that matter, even Texas wanted help to build a dam on the Canadian.

The construction of medium to large dams remained the provenance of the experts at the Army Corps of Engineers or the Bureau of Reclamation despite efforts by individual civil engineers in the west. Over time the two agencies worked out a rough division: pure flood-control dams and channel-control structures (weirs) belonged to the Army Corps, while irrigation and power generation came under the Bureau of Reclamation's oversight. That said, during the 1920s and 1930s a great deal of interdepartmental strife remained between the Bureau and the Corps. Adding to the jurisdictional confusion, the Department of Agriculture waded into the fray from time to time.[14]

The Arkansas Basin states wanted federal assistance with construction of their flood-control dams and lobbied for it. They argued that flood-control projects in the individual states would benefit the nation as a whole because floods caused losses that affected others through infrastructure repair costs, lost time in transit, higher costs for food and fiber due to flood-related crop losses, and damage to navigation channels on larger streams. Flood control arguments became louder after April 1927, when the wet spring and winter in the upper Mississippi River Valley and along the Arkansas and lower Missouri led to a surplus of water in the Mississippi River watershed. The Mississippi began rising, and rising, and rising, breaking levees and inundating thousands of square miles of land in Arkansas, Mississippi, and Louisiana and taking hundreds of lives in the Mississippi Delta south of Memphis, Tennessee. As the waters slowly receded and the almost unbelievable costs to the country became apparent, the upstream states pressed their arguments again. If more of the water had been held back in Kansas, Arkansas, and Oklahoma, the flooding elsewhere would not have been so bad. A dam in New Mexico would benefit New Orleans, the Arkansas River Basin Association argued. That it would also benefit New Mexico remained unsaid.[15]

All the commotion and pleas for flood control downstream failed to answer the question of which federal entity should build the dams in Texas and New Mexico. The Army Corps "did" flood control because of doing river-channel and navigation protection, so they were the logical choice. Army engineers also had a great deal of experience and a good record of sturdy dams. However, as per the orders of Congress, the Army Corps focused on maintaining the navigability of streams, and the Canadian was navigable only downstream of central Oklahoma. The Army Corps's mandate also required that the cost of prevented flood damage had to exceed the cost of the dam. This would be difficult to prove, given how unpredictable the Canadian's floods were and how infrequently truly large inundations occurred. The flood-damage clause would return to haunt Texas dam proponents, and even in the 1920s it caused some concern. If Amarillo, Texas, or Tucumcari, New Mexico, had been built in the Breaks along the banks of the Canadian, life would have been easier for Sid Stinnett, Lee Bivins, Arch Hurley, and their supporters. However, another option existed.[16]

"Conserving" the wasted waters in order to "reclaim" barren land fell to the Bureau of Reclamation, the other great dam builder in the west. The Bureau sprang to life in 1904 under the Newlands Act, named for Senator Francis G. Newlands of Nevada, and had the stated goal of bringing unused water to where it was needed, reclaiming millions (if possible) of acres of fertile but dry western land for the use of family farmers. If the water captured by a project supplied an irrigation system, then the Bureau of Reclamation was available. They built both the necessary dams and the irrigation canals, offered financing paid for by the water user, did not require proof of cost/benefit based on downstream damage, and had experience working in New Mexico. But the results of Bureau of

Reclamation projects in the state thus far failed to please New Mexico's state engineer. The Bureau also limited the amount of land participants in a Bureau project could own as a way to ensure that developers and corporate farmers did not hog all the benefits of the project. The ranchers and dry-land farmers in Texas and New Mexico held far more than the Bureau's maximum of 160 acres and would have to sell at the lower dry-land prices in order to accept water from any project. Even worse for the interests in Amarillo could not work with the Bureau. The goal of the Bureau was to encourage settlement and development of *federal* lands, of which Texas had none. Individuals either leased Panhandle farmlands from the state or owned the land in fee simple. The entire Canadian Valley belonged to ranchers, including Lee and Julian Bivins, R. B. Masterson, and others. Texas municipalities and irrigators confronted the disadvantage of the state's enormous land holdings and began searching for a way around them.[17]

To complicate things further for the Panhandle conservationists, neither the Bureau nor the Army Corps built municipal water projects, leaving Amarillo and Borger literally high and dry. Although municipal diversions from some Bureau projects existed, they were purely incidental to the irrigation aspect of the project, much in the same way that any electric power generated by the Bureau dams was a side benefit, not the Bureau's goal. As the prospects of obtaining a federally funded dam grew more problematic, dam proponents in Amarillo and Borger no doubt wished that they could find a way to obtain the dam without federal involvement. But the cities could not afford to hire a private company to build a dam, install pumps, and hook the river into the city systems. There had to be a way to finance a Canadian River project, but how?[18]

All hope of federal help faded away in September and October 1929. Neither the Bureau nor the Army Corps of Engineers had funds for small dams after the economic crash and the ripple effects that followed. In Texas and New Mexico, falling commodity prices collided with declining rainfall as an eight-year drought slowly took hold of the High Plains. The relatively good times of the 1920s came to an end, and the region soon became the focus of federal attention and even international attention, but not for the reasons area boosters wanted. Local eyes would turn to the Department of Agriculture, to Borger's city manager, and to an eternally optimistic newspaper editor as the unlikely saviors of the High Plains' water.

Chapter 3

Saving Water to Save the Soil

The 1890 report of the *Tascosa Pioneer* would have sounded terribly familiar to Amarillo residents reading it forty years later:

Wednesday was a windy enough day, but it was about midnight or a little after that things went to shaking and nodding in the breezes and getting a moving on themselves about right. From that until daylight was the hardest windstorm the oldest inhabitant knows anything of. It just naturally blew and blew and blew, and blowed and blowed and blowed, and swept the country all up in one great big continuous sweep. In the latter half of the night it piled dust heaps everywhere, and sent it through and into the tightest buildings, and rattled the roofs and shook the fences and scattered loose boards and boxes and barrels and bent the trees and roared and howled and shrieked and hissed till nothing else could be heard.[1]

The small cow town newspaper described the onslaught of one of the 1890 drought's more memorable dust storms very well. High wind picked up the drought-loosened soil and the river-valley sand and carried them through any opening, no matter how small. The air smelled of dust and ozone, of static electricity generated by the rolling soil. Dust settled onto everything: clothes, tables, food left out under fly-screens, the bodies and chimneys of kerosene lamps, the people in bed. The next day would be cleaning day as the householders and business owners tried to send the layer of sand and soil back outdoors where it belonged, hoping that it would be a very long time before anything like *that* happened again. March 1890 was going out like a lion, and the residents of Tascosa no doubt hoped that the rains would come in as the month departed.

Forty years later, other dust storms brought the Canadian River Valley and surrounding High Plains to national attention, but it was not the sort of attention residents and area proponents wanted. Some expert observers believed that poor land-use caused the collapse of dry-land farming, many ranches, and the regional economy, creating a textbook example of the evils of unscientific land management. Long-time area residents knew that, to paraphrase the prayer book, "drought has come, drought has gone, and drought will come again," along with dust clouds and starving bovines. The semiarid environment of the High Plains shifted to truly arid at least once a decade for a year or more, making drought a familiar and unwelcome event. Was the drought of the 1930s really out of the ordinary, or was it the number of witnesses and amount of documentation that made it unusual?

The differences came from observation, duration, and response. Far more people witnessed and studied the drought of the 1930s than had commented on the dusty 1850s. The meteorological drought also lasted longer than other dry episodes between 1890 and 1930. But the greatest differences between the drought of the 1930s and those of the mid 1920s, the 1890s, or the 1850s stemmed from federal interventions. For the first time the federal government became directly involved in drought relief. During the 1920s High Plains water developers consulted with agencies of the federal government, but after 1932 federal money and federal experts swept through the region bearing plans for improvements. As the decade progressed, local residents developed a push-pull relationship with federal largess that continued through the rest of the century: appreciative of the funds and of some assistance but irritated with and at times resentful of the orders and advice from "outsiders," "Easterners," and bureaucrats who seemed unwilling to listen to and learn from local experience. Like other westerners, conservationists from the High Plains pushed back, trying to steer federal efforts to aid local ends.

Although the 1930s brought the Great Plains' worst drought (or drouth) in memory and greatly increased the federal presence in the Panhandle and South Plains, local residents drove many developments and plans to alleviate the unceasing dryness "from below" by determining what type of water conservation projects were needed and where they should be, then working with representatives of the federal agencies to plan and build the projects. The combination of the Great Depression and an intense multiyear drought also led to local and national calls for conservation. Local conservation proponents, including newspaper editor and regional organizer John McCarty of Dalhart and Amarillo, argued that if only the "wasted" waters could be saved, then the soil would stay put and grasses (including wheat) would reappear. The return of the grasses might in turn lead to the return of prosperity lost after 1929.

The collapse of prices on the New York Stock Exchange on October 29, 1929, marked an international financial "panic" and a chain of events that led to economic depression. Even if they were not active in stock-market speculations, farmers and ranchers in the High Plains felt the Depression's effects before the drought began. The combination of overproduction, international economic woes, and British and European reciprocation for the Smoot-Hawley tariff brought about a commodity price collapse that the drought of 1933–39 only made worse. As discussed in chapter 1, the rapid advance of mechanized agriculture, specifically the surge in dry-land wheat farming that historians later called the "great plow up," led to larger and larger harvests of wheat and other commodities. Grain prices sank again as property taxes remained stable and the cost of equipment and fuel rose. Federal farm assistance faded away at the same time that world commodity prices began declining in 1930. Europeans would not have the money to pay for wheat and other products from the United States even if the governments of Great Britain, France, Germany, and other nations had not erected "tariff walls" in response to the Smoot-Hawley protective tariff. Suitcase farmers and farm tenants who rented the land they plowed and who lived from harvest to harvest were some of the first to feel the effects of both the Depression and the drought. The solution for individual farmers seemed obvious— plant more.[2]

And so in the spring of 1930 and 1931 tractors chugged across the plains once again. The Farm Bureau and other groups lobbied for the McNary-Haugen bills, but even if Herbert Hoover had been inclined to support them (which he remained reluctant to do for constitutional and philosophical reasons), the federal government had no money to pay for the crops or to offer credit to farmers. Two good harvests, especially the 1931 wheat harvest, sat heaped up on the ground, waiting for trains, or piled up in grain elevators because no one was buying wheat, corn, rye, oats, and sorghum. Cattle prices also declined, as did the wool market. At first the Panhandle seemed to hold on to prosperity a bit longer, protected by oil income and the jobs that the carbon-black plants and refineries provided. Then the East Texas field came into production, and oil joined other products of the land in a price free fall as oil producers mimicked farmers by pumping all that they could, although for different reasons. It was in the midst of this bounty that drought returned to the High Plains.[3]

Drought and dust storms gripped the High Plains and South Plains as they had before. As the agricultural historian James Malin wrote in a series of articles looking at dust in Kansas before the twentieth century, multiple newspaper articles from the 1820s through the 1850s described episodes of blowing dust. Some of that blowing topsoil quite probably came from what would become Texas, New Mexico, and Oklahoma, although reporters and observers of the time had no way of tracking the source of the storms. Mrs. Minnie Hobart told historian L. F. Sheffy that she arrived in the Texas Panhandle from Vermont in October 1888 and that night encountered the first of many dust storms that she was to witness in the region. Tascosa experienced blowing dust and sand in the 1890s, and accounts of early Hutchinson and Moore county history include descriptions of at least one black duster in 1903. W. J. Morton, a farmer in northeastern Moore County, described one day in March 1903 with a "black duster from the north," after a hard winter. "It was one of the biggest and blackest dust storms I have ever seen and there was scarcely any land being farmed at the time." On May 2, 1904, dust flew so thickly in the Oklahoma Panhandle that a train was forced to wait in Texhoma for twelve hours before it could move safely. Far to the west, where the Canadian joined the Conchas in New Mexico, beyond the great plow up, the weather reporter on the Bell Ranch noted dust and sand storms in 1899 and 1901, as well as in May 1927 and the 1930s. Even in 1935, the editor of the *Panhandle (Texas) Herald* noted in passing that the massive dust roller of April 14, 1935, lacked the aesthetic qualities of the March 17, 1923, storm: the former was "not as resplendent in color effects" as the latter. As historical geographer Geoff Cunfer notes, the land blew without being plowed or heavily overgrazed, if the soil was dry enough and the wind strong enough. So what made the drought of the Dust Bowl different?[4]

The expanse of bare ground covered in loose soil was one difference. The dry-farming techniques favored in the High Plains depended on creating a "dust mulch" using a one-way disk plow. After deep plowing the soil and planting the grain with a seed drill or other planter, farmers went back and turned the soil again. This created a light, airy layer of dust intended to keep the water in the soil by blocking evaporation. After each rain a good husbandman went out and plowed under the wet soil, turning the moisture in so that it would not evaporate. The one-way disk plow caused problems during the extreme dryness of the mid-1930s. As wheat farmer W. J. Morton described it, the one-way's ten-foot-wide row of vertical cutting disks shaved off the grass and powdered the soil. "Before we knew it we had one-wayed tons and tons of topsoil into the air. The least little old wind would start the dirt to sifting. In plowing we just moved the dirt over to one side and then turned around and moved it back. By the time the thirties arrived, our farm land was in a real moving condition and boy, did it move!" Most of the region's moisture came in spring and early summer, leading to a double plow problem: farmers planted winter wheat in the early fall only to watch soil blow in the dry autumn if no rain came. Then they plowed in the spring to try and get a sorghum crop in if the wheat failed. The high winds of March and April carried off soil and seed both. Good farmers also conserved moisture using summer fallow, plowing in any weeds that might try to sprout and steal the moisture. A clean field was a sign of good husbandry until the winds came and the rain failed.[5]

Although timely rains in the High Plains led to a record dry-land crop in 1931, hardships began soon after. A dry season and the effects of the Depression had taken their toll in 1930, but farmers planted anyway—they needed cash, and drought crops such as broomcorn and sorghum brought in less income than wheat did. Heavy snows that winter had killed people and livestock, but the moisture put a good season into the ground. The spring drought hurt yields a little, but the dryness did not come until late spring, after the wheat was already up and almost mature. The season proved to be so good that it set records for Texas, Kansas, and Oklahoma dry-land grain yields. But the grain sat unsold, and farmers got as little as twenty-five cents per bushel. Even the next year, enough surplus wheat remained that buyers gave only thirty cents per bushel for that year's crop, such as it was. For those who rented the land, and even those who owned, it seemed better to wait out the next year on their savings and surplus.[6]

As a result, tens of thousands of empty acres lay bare to the sun, carefully tilled to keep out the water-stealing weeds. Instead, the winds came. What wheat had been sprouting died under the hot April wind. Without moisture to help hold the soil particles together, the moving air began shifting the soil, first into small drifts, then lifting the dust and turning the air a brassy tan before erasing the sky with a "black roller." When dry cold fronts blew in, as was common, they caused some of the worst dust storms as cold air collided with hot or warmer surface air. Static electricity generated by friction within the moving dust attracted more dust particles, adding to the darkness. Topsoil that had required centuries to develop under the short grasses of the semiarid High Plains disappeared eastward, never to return. As the dry months crept on, drought-powdered pastures began disappearing along with the plowed ground.[7]

At first, ranchers suffered less than did the wheat growers. Ranches such as the Matador's Alamositas division, the Bell, the LX's Bivins branch, and others with live water fared all right for a while. Cattlemen tried to reduce the number of cattle on the land in order to keep from ruining the grass, but the cattle market softened due to low demand, and as a result even animals in the best condition commanded low prices. Stock farmers who could raise irrigated garden truck, such as the Kohler family in Cimarron County, Oklahoma, up on the Dry Cimarron River, were better off than monocrop grain producers because the family had enough to eat even if cash was scarce. A. P. Atkins, who owned 10,000 acres (4,050 ha) in the central Oklahoma and northern Texas Panhandles, reported that he managed to salvage something by rotating his summer pastures as well as his winter ones, feeding a little from bottomland alfalfa and winter wheat, and cutting his herd from 500 head to 250. Other stockmen suffered along with their farming neighbors when drought withered their garden truck and blowing sand and dust cut off young plants or buried them in drifts of soil. As the drought deepened and continued, everyone felt the pain even where the dust remained in place. Where the soil flew, so did the prospects for the next year.[8]

By 1935 parts of the High and South Plains had become environmental disaster areas. Those farmers and ranchers whose land remained in place confronted drifts of their neighbor's soil that covered the grass and smothered any plants coming out of the ground. The dust coating the grass ground down the teeth of cattle, sheep, and horses and made the forage unpalatable. Skinny cattle with xylophone ribs became all too common a sight and one that would be recorded in photographs, sketches, and paintings for the rest of the nation to see. Furthermore, when the rich topsoil disappeared eastward, the remaining hard ground lacked loam, was less amenable to the plow, and absorbed less of what rain or snow did fall. As a result, water erosion deepened gullies and arroyos and caused flash floods downstream. The agronomist Paul Sears described the process as one of "deserts on the march." Adding insult to injury, some individuals found that the very government programs instituted to try to help farmers and ranchers led to further hardship.[9]

The story of Franklin Delano Roosevelt's election in 1932 is well known, as are the numerous programs that he initiated during the first one hundred days of his presidency in 1933. One of those programs, a part of the original Agricultural Adjustment Act (AAA), was directed at cotton farmers. The AAA's crop-reduction program attempted to raise the prices of some commodities by paying farmers to plant less or to plow under what had already been planted in order to reduce the surplus. Farmers did not have to participate, but those who did received better prices for their product, and a number of cotton farmers took the government up on its offer, often taking rental land out of production first and evicting their tenant farmers. The ongoing drought "helped" by reducing cotton harvests on the South Plains even without federal incentives. The combination of drought and payments raised the price of cotton, a wonderful development for any cotton farmers who had made a crop. However, the process hurt the now unemployed and homeless tenants and the cattlemen who needed to buy cottonseed meal.[10]

The declining cotton crop created a near-nightmare for some ranchers because increasing the price of cotton also raised the price of cottonseed meal, the high-energy drought and winter fodder that cattlemen had turned to in order to spare at least part of their range. The rancher and historian J. Evetts Haley penned an eloquent and unhappy description of the program and its results entitled "Cow Business and Monkey Business" for the December 8, 1934, *Saturday Evening Post*. In it he described how a veterinarian came to the Haley family's Permian Basin ranch in order to sort and then shoot cattle for the government's cattle purchase program, killing any animals too weak or too ill to use for food. Haley's article pins the cause of the cattle's deaths both on the federal government's crop reduction program and on federal assistance paid to those ranchers who should have failed in the first place. There was no cottonseed cake, or if there were any it was too expensive for a rancher to buy. Haley reminded his readers that drought had always forced the most foolish and ill-prepared out of the cow business in the past, so why should this dry spell be any different? According to Haley, his family's cattle could have survived the drought except for Washington's "monkey business," which included manipulating cattle prices. Haley also argued that the federal government punished ranchers for their slowness in participating in the AAA by forcing the ranchers into contracts with changeable requirements and restrictions. Meanwhile, as J. Evetts Haley and a few others protested the AAA, other federal agencies introduced new programs into the High Plains, often at the instigation of local conservationists.[11]

John L. McCarty, historian, newspaper editor, and son of a tenant farmer, believed in conservation of people and of the soil. Born in 1901, McCarty moved to different ranches and farms around the Panhandle over the course of his childhood as his father, William Rush McCarty, worked at a variety of jobs. The family lived briefly in Casimero Romero's home at Old Tascosa, a residence that had a lasting effect on McCarty's later interests. As McCarty later put it, "I suddenly realized that I was living in history. History surrounded me and whatever happened around me would one day be history." During the course of his career as a newspaper writer and editor, John McCarty interviewed hundreds of old-time Panhandle residents—ranchers and farmers, lawmen and others—and became very much aware of the continuity and, as his biographer put it, of "[the] 'spirit' of the past." He was an agrarian, a member of the Farmers Union, and someone very much interested in keeping alive, preserving, and conserving the best of what he thought to be the special spirit of place in the High Plains.[12]

In 1929 Eugene "Gene" Howe, the owner of the *Amarillo Daily News* and of several other smaller local newspapers, sent John McCarty to Dalhart to take over running the *Dalhart Texan*. It was from here that McCarty first made a name as a writer and regional booster, and it was here that he would develop the interest in conservation that led him to spend the latter half of the decade working for water storage projects in the High Plains. The year before, McCarty had interviewed A. S. Stinnett at length for an article in the *Texas Commercial News* about flood control and water conservation, and there is little doubt that the seed planted in the course of the interviews grew to maturity during the Dirty Thirties as McCarty did what he could to try to help keep Dalhart alive in the middle of the Dust Bowl.[13]

From the local onset of the Depression and drought in May 1932 until Howe brought McCarty back to Amarillo in 1936 following the sudden and unexpected death of the *Daily News*'s manager, John McCarty was the voice of Dalhart and the northwest Panhandle, advocating conservation, patience, and perseverance. In 1932 McCarty and several other local businessmen and ranchers funded efforts to dam Rita Blanca Creek, a spring-fed stream just south of town, in order to develop a park there. For the next two years McCarty took a calculated risk and wrote editorials critiquing some of the current popular farming and ranching practices, pointing out that what worked for one individual could spell disaster when everyone did the same thing. He cited leading soil and water conservation experts such as Hugh Hammond Bennett, who encouraged leaving crop stubble to hold the soil and deep plowing in the fall as well as contour plowing. In August 1934 McCarty invited soil conservation specialist H. H. Fennell from the Dalhart Wind Erosion Control Project to write a series of articles about conservation techniques for the *Daily Texan*. At the same time, McCarty and Fennell argued against the Relocation Administration, Secretary of the Interior Harold Ickes, and others

Figure 6: Start spreadin' the news: Newspaperman and Texas Panhandle promoter John McCarty at his desk in Amarillo. Used with permission of the Panhandle-Plains Historical Museum, Canyon, Tex.

who urged farmers and ranchers to abandon the High Plains entirely. McCarty editorialized that the residents of the High Plains possessed the determination and capability to carry on and prosper if they had the right tools and training. One effort to encourage Dalhart and area residents to stick out the drought came in 1935: McCarty founded the "Last Man Club" for those who pledged to stay and keep the area going. His rebuttals of the popular media accounts of life in the Dust Bowl brought McCarty and the *Texan* national recognition, but also led to charges of hypocrisy and betrayal when the founder of the "Last Man Club" left Dalhart in 1936.[14]

McCarty's recall to Amarillo to run the *Daily News* did not stop his conservation efforts. Because of his reputation as a regional booster as well as a conservationist, members of the Chambers of Commerce of Canyon and Hereford approached him when they wanted to obtain federal or state assistance to build a reservoir on Tierra Blanca Creek. The previous year W. A. Warren of the Canyon Chamber of Commerce had called for damming the creek, a call supported by Gene Howe of the *Daily News* and by Randall County Commissioner George Heath. One flash flood too many had gone down the stream and through Canyon: the time had come to stop the water. According to conservation proponents, Tierra Blanca Creek needed to be dammed in order to reduce flooding and to provide a place for boating and fishing, the lack of which Howe had been lamenting for a decade. But little happened because the project was too small for the Army Corps to deal with (they were busy damming the Canadian in New Mexico), and the Bureau of Reclamation did not see any irrigation prospects in the project.[15]

The Canyon businessmen continued trying, and after talking with their counterparts in Hereford, the men of the Chambers of Commerce approached McCarty. Rather than limit the discussion to just residents of Canyon and Hereford, the conservationists organized a meeting in Amarillo and announced it in the Amarillo and other papers. Forty-eight people attended the December eighth meeting. Editor Gene Howe described the need for more recreational opportunities, including fishing and boating and waterfowl hunting, while W. H. Fuqua, an Amarillo farmer and developer, discussed contour plowing and other techniques. It is critical to note that men from Vaughn and Clovis, New Mexico, also attended and addressed the gathering, pointing out how this was a matter for regional interests and efforts. As New Mexican K. C. Lea put it, Clovis and Curry County were "too far east to be in New Mexico and too far west to be in Texas" but they shared the Panhandle's topography and climate. The men decided to hold a second meeting of the tentatively named "Panhandle Conservation Association" in order to get more participation from those in the surrounding area. County commissioners, judges, and interested landowners received invitations, and meeting organizers also encouraged representatives of the Soil Conservation Service, Works Progress Administration (WPA), and Resettlement Administration to attend.[16]

To the great surprise and pleasure of the conservationists, almost five hundred Panhandle farmers, ranchers, businessmen, and civic leaders descended on the Amarillo Hotel on December 19, 1936. Definite interest existed for the ideas proposed by the residents of Canyon and Hereford, to put it mildly. Two major themes emerged at both meetings: irritation with the lack of federal attention to local suggestions and unhappiness with plans proposed by Rexford Tugwell and the Resettlement Administration to purchase or simply take back the formerly federal land in Oklahoma and New Mexico and move people out of the region. Local residents wanted to stay, and they had ideas for conservation projects that would provide jobs, protect the water and soil, and open up new recreational opportunities. After some discussion, the group formed the regional Panhandle Water Conservation Association (PWCA). John L. McCarty was elected president and Carl Hinton of the Amarillo Chamber of Commerce became the secretary and legislative representative. Dues came from each county and would be used to pay for postage and for Hinton to go to Washington, D.C., and to lobby for projects. This entire organization and its goals fit in very well with McCarty's earlier work because the group wanted to save the area's soil by saving the water that kept the soil in place.[17]

The conservation interests, regional emphasis, and interstate cooperation advanced by the PWCA provide a good lens for looking at the larger regional and national picture. Amarillo organizations and businesses had already begun working across state lines by assisting the executive branch of the State of New Mexico. Through a combination of persistence, arm-twisting, personal contacts, and appeals to regional mutuality, New Mexico had obtained a dam on the Canadian River at the Conchas–Canadian confluence. As early as 1924 the office of the New Mexico state engineer proposed building a dam at that location, but the state had no money available at that time. The tristate Canadian River Compact offered some hope for assistance until October 1929, when funding dried up like the soil of eastern New Mexico. Once the New Deal relief efforts reached the state, it appeared to eastern New Mexicans that the more heavily populated Rio Grande watershed and the Navajo and Ute Indian reservations in the western part of the state received all of New Mexico's Civilian Conservation Corps, WPA, and other projects. Secretary Ickes himself objected to a dam on the Canadian because of his dislike of New Mexico's governor, but persistence and Governor Clyde Tingley's personal friendship with President Roosevelt carried the day for a dam, provided that New Mexico paid for the land under the dam (since the site was not on federal acreage).[18]

Governor Tingley appealed to New Mexicans and Texans alike in his efforts to purchase the land needed for the construction right-of-way and dam access. Seventy-nine of Amarillo's businesses stepped up to the challenge, sending amounts ranging from five dollars from Cunningham Floral to five hundred dollars from the International Harvester dealership and three hundred from Amarillo's *Globe-News* publishing corporation, for a total of six thousand dollars, in order to support a dam that would in turn support the regional economy. The Conchas Project would be a flood-control dam built and operated by the Army Corps of Engineers, constructed with both contractor and WPA labor. The project also involved the Texas Panhandle regional WPA director, A. A. Meredith of Amarillo, to a limited extent.[19]

Outside the High Plains, people remembered 1935 for high water and drought. Strong summer storms in May and August in northeastern New Mexico, western Nebraska, and southeastern Colorado dumped much-needed rain too quickly for the dry, bare ground to absorb. Floods roared down the South Platte, Republican, and Arkansas Rivers while normally dry washes and streambeds in the southeastern corner of Colorado filled to the brim and carried more of the drought-loosened soil into Oklahoma. Without plants to slow the water's flow over the ground or to help the soil absorb the moisture, most of what fell was "squandered" downstream. The conservation-minded men and women of the High Plains must have read the news reports and wondered why all that water was being allowed to go to waste. But things were changing.[20]

According to historian Vance Johnson, even the federal soil conservation experts became frustrated with the lack of communication and coordination. So much could be done, and yet great clouds of soil continued rolling eastward, first from the southern plains, then from Montana and the Dakotas, then from the south again as the drought intensified or moderated, shifting north and south. Johnson described a conference held on January 18, 1936, that included more than twenty men who "represented half a dozen state and federal farm agencies, and among them were some of the best scientific minds of the southern Plains. All were experts on soil and its control." They had met the month before in Denver and now came together to coordinate the collection of data on every acre of blowing land in the High Plains and to plan how to use the $3.5 million they had requested from the Department of Agriculture. The funds proved to be unavailable, so the group drew up a resolution urging that federal funds be released to pay farmers to chisel plow, contour plow, and otherwise put into place the best-known soil conservation practices on a permanent basis. Congress added the experts' ideas to the planned Soil Conservation and Domestic Allotment Act. President Roosevelt signed the act into law on February 29, making possible the programs that so interested the members of the PWCA later that year.[21]

Nineteen thirty-seven proved to be a watershed year for regional conservation efforts. The still-uncompleted Conchas Dam, the flood control and irrigation project at the Canadian and Conchas Rivers junction that had been authorized in 1935 but not begun until more than a year later, caught floodwaters from the spring melt and heavy rains in May, filling Conchas Reservoir much sooner than anticipated. Despite the new dam's presence, flooding downstream cut Tucumcari off from the outside world by washing out roads and railroad bridges. High water in the Panhandle carried off more precious topsoil and drowned several people caught in low places or while crossing flooded streams. The flooding reminded regional conservationists that time was of the essence in stopping the "wasted water" and its load of topsoil.[22]

The PWCA received the recognition of the Arkansas Valley Association, a regional conservation body, at the same time that the Dust Bowl legislators were forming a regional body much like the older Farm Bloc. In May of 1937 the PWCA sent Carl Hinton to Washington, D.C., posthaste with instructions to make connections and to work on obtaining support for the first four dams: Tierra Blanca Creek, Rita Blanca Creek, McClellan Creek, and a fourth in the northeastern Panhandle. Overspreading all this was the passage of the Bankhead-Jones Farm Tenancy Act, a modestly titled piece of legislation that brought the dreams of fishermen in the Panhandle and eastern New Mexico closer to fruition.[23]

Before it could do anything toward directing federal moneys, the PWCA needed legal designation as a chartered organization in order to collect and use dues to lobby, and to obtain recognition from politicians and agency heads in Washington as well as from other regional organizations. The Southwest Valley Association, a conservation group that included the states in the Arkansas River Basin, had recognized the PWCA at the SVA's February meeting, and John McCarty had been named as a vice president of the SVA. Back home, Judge Wilson Cowen of Dalhart prepared a bill for the Texas legislature naming the PWCA as a body that "could effect water and soil conservation, irrigation and flood control projects, reclamation, and recreational development for the basins of the Brazos, Canadian and Red Rivers." Cowen finished the proposed bill on February 24, 1937, and State Senator Clinton Smalls introduced it into the Senate. The bill passed, and Governor James V. Allred signed it into law on May 4, 1937. The Panhandle Water Conservation Association now had legal standing and could function as an agency within the thirty-two member counties, negotiating with other state and federal agencies in order to meet the conservation needs of the Panhandle and adjacent counties.[24]

McCarty, Carl Hinton, and the others wasted no time and soon called another large meeting to order on May 20, 1937. Representatives from Texas, Oklahoma, New Mexico, Colorado, and Kansas attended the gathering, hosted by the PWCA and Amarillo's Chamber of Commerce. Earlier, outgoing Governor Tingley of New Mexico had issued an enabling act, permitting those counties interested in the PWCA's programs to participate with the blessing of the state engineer's office, an act that newly elected governor Andrew Hockenhull fully supported. B. W. McGinnis, the assistant conservationist for the Soil Conservation Service of the Department of Agriculture, pointed out the PWCA and the Soil Conservation Service shared the same goal: "holding water where it falls in order to force it into the ground that it may help to produce more vegetation" and thus hold the soil. He argued for check dams on washes and arroyos as one way to keep water at home.[25]

In his follow-up address, John McCarty described the composition of the group and said that he had received a phone call that morning from a "gentleman who feels his convictions strongly on matters of this kind" and who had wondered exactly what sort of people would be at the meeting. The PWCA's members included mostly "dirt farmers" and ranchers but also nineteen county agents, thirty-four other government officials, some lawyers, businessmen, twelve bankers, nine newspaper men, and sixty county commissioners, according to the rolls. Following McCarty's talk, soil specialist H. L. Hauter listed the causes of the area's problems: lack of rain, plowing land best left as pasture, improper crops, and land in uneconomical units (i.e., farms and ranches too small for dry-land or too large for irrigation). His solution was for landowners to get the latest facts about soil conservation and crops and to practice soil conservation techniques, and for the various state and federal agencies to work together.[26]

After lunch the PWCA membership drew up five resolutions: first, that a five-state conservation authority or planning board be formed; second, that federal agencies needed to coordinate better—the Soil Conservation Service and WPA did not always plan together; third, that the federal government needed to add more local interest projects to President Roosevelt's water conservation and flood control program. The fourth resolution stated that Carl Hinton be sent to Washington seeking federal approval for small dams and reservoirs, and the fifth, that these projects be added to Nebraska Senator George Norris's flood control bill or for Representative Marvin Jones of Texas to present them in a separate House bill.[27]

Carl Hinton, the PWCA representative to Washington, spoke after the group voted to adopt the resolutions. It was evident that he had been very busy meeting with senators, representatives, and members of various federal agencies in his efforts to bring the needs of the High Plains to the attention of those in power. He began by explaining that even though many in the room did not care for the term "Dust Bowl" and wanted him to find another, President Roosevelt and others used the phrase even if it was not "necessarily good or true," and so Hinton would continue saying "Dust Bowl." Hinton then gave a little history of the PWCA and reminded the listeners that he had gone to Washington, D.C., as soon after the December 19 meeting as possible.[28]

Over the course of his trips Hinton met with legislators and organized a conference of representatives and senators from Dust Bowl states. They in turn had formed a permanent committee despite the opposition of Roosevelt and Secretary of Agriculture Henry A. Wallace . Another difficulty had come from Senator Norris, whose flood control bill divided the United States into eight sections, putting part of the Dust Bowl in with the Gulf Coast. One imagines head shaking and murmurs about politicians' lack of geographical knowledge before Hinton continued, explaining that he had opposed the Norris bill because all the funds were designated for "building reservoirs, dikes and dams on the lower Arkansas, lower Red and lower Mississippi Rivers" instead of the smaller headwaters conservation projects envisioned by the Dust Bowl Committee. Hinton had protested and was told that the flood problems had to be fixed first and then they could get around to "your" problem. As Hinton pointed out, downstream legislators "don't think in terms of water conservation," and he allowed that "if you lived [on the Mississippi] you wouldn't either." Adding insult to irritation, the rather pugnacious and blunt (and powerful) Representative Clarence Cannon of Missouri did not want to do anything to encourage people to continue living in the High Plains. They should move "into regions where it does rain." The minutes did not record the audience's response to the honorable gentleman from Missouri.[29]

The Norris bill, U.S. Senate Doc. 2555, promoted flood control and conservation but failed to address efforts to stop wind erosion, even after graphic evidence of the problem drifted down over Washington. The Norris bill encouraged the PWCA to work harder for local control over conservation plans and projects and served to confirm what some High Plains residents had come to think about federal inefficiency. Federal erosion control specialists, including Hugh H. Bennett, worried about water erosion, not aeolian (wind) erosion. For example, the speaker at Dalhart, H. H. Fennell, initially focused on terracing to prevent rainfall runoff. The 1933 Agricultural Adjustment Act made no provision for efforts to stop wind erosion, all of which demonstrated from just how far "behind the curve" the federal conservationists had to come. One wonders if Hinton, McCarty, or the senators from New Mexico had been tempted to enquire of Norris, "would it not be less expensive and more efficient to stop the flood waters *before* they reached the lower Mississippi and Arkansas?" But it was already becoming apparent that conservation of funds was not the conservation that most interested FDR, Secretary Wallace, and other New Dealers. Their initial opposition to a Dust Bowl states conservation region helps illustrate how the New Deal's experts viewed the problem, what solutions most appealed to them, and why this irritated regional residents.[30]

Although conservation held the attention of the United States in a way not really seen since the Theodore Roosevelt administration, President Roosevelt and his advisors and experts, including H. H. Bennett, Rexford Tugwell, Secretary of Agriculture Henry A. Wallace, and others, had very mixed emotions about local conservation groups such as the PWCA. While federal officials encouraged local suggestions for federal projects, the New Dealers were less happy about the rise of independent regional blocs. In 1936 a group of farmers banded together as the Southwest Agricultural Association, with members from Texas, Colorado, Kansas, Oklahoma, and New Mexico. In April of the next year the Farm Practices and Legislative Committee of the SAA drew up a list of resolutions, including "[a federal] authority set up with in the confines of the Dust Bowl" to coordinate everything. The farmers also recommended a declaration of martial law in order to force people to take care of their land and stop it blowing, neither of which was quite what the various federal agencies already working in the area had in mind. As Carl Hinton explained, even the March 1937 formation of a legislative Dust Bowl Committee made President Roosevelt and Secretary Wallace concerned about the potential creation of "a little TVA" that could challenge federal authority and control over regional plans and projects.[31]

As historian-geographer Geoff Cunfer and historian Paul Bonnifield point out in their work, these men already had plans for the High Plains, and they did not necessarily include keeping the farmers on land that seemed "sub-optimal" for farming. It struck the New Dealers as inefficient and even irresponsible not to find ways to remove people from the blown-out, ruined, and barren landscape of Cimarron, Dallam, and Seward Counties, to name three of the more infamous locations. The 1934 Taylor Grazing Act provides one example of this: it terminated homesteading on the remaining unclaimed federal lands and favored government-regulated ranching over farming in the grasslands of the plains. As Wallace and some others saw matters, trained technical experts should analyze regional suggestions for conservation and land use, assuming the ideas did not conflict with current federal plans. If the proposals seemed workable then experts who knew the best scientific and engineering methods and practices should implement the projects, controlling them from above. This centralizing mindset upset old Democrats, including J. Evetts Haley, and contrasted with the bottom-up ideas of John McCarty, A. S. Stinnett, and other Panhandle conservationists. They wanted federal assistance and advice but on their own terms, as a way to help keep people and communities in the High Plains. These philosophical differences led to frustration that continued as long as federal agencies remained actively involved in water conservation projects in the Panhandle. The "comprehensive land-use planning" and federally directed crop production and marketing dreams of the most progressive of the New Deal agricultural reformers were rather at odds with the desires of many residents of the Canadian River watershed; but in April 1937, no one spoke of this publicly. Instead, John McCarty, Texas representative Marvin Jones, and others took a wetter spring and the official recognition of the PWCA as encouragement for their efforts.[32]

To go a little further, it is possible to sum up some of the philosophical differences between McCarty, Marvin Jones, and other Dust Bowl "stayers" and some of the New Dealers as stemming from contrasting ideas about efficiency and scale. As early as 1923 Lewis C. Gray and others had outlined the desperate need for a neatly ordered, well-planned system of land use based on such considerations as climate, industrial potential, land type, agricultural technologies, culture, and local history in a piece entitled "The Utilization of Our Land for Crops, Pastures and Forests" in the *Yearbook of Agriculture*. Without this most-efficient use of land, Gray argued, the country could not support a projected United States population of 150 million. The depopulation of the High Plains would be one result of this more productive, federally controlled land-use plan as the national government converted farms into large ranches or larger and more efficient farms, while the region's former residents moved to areas with more sustainable farming and industries. Farming the High Plains as done at the time was not efficient or desirable for the nation as a whole, according to the Gray study.[33]

A number of residents of the Dust Bowl region disagreed with this assessment and strongly resisted suggestions that they move. They also disagreed with the suggestion that the disaster was entirely or even mostly their own fault and growled at the implication that farmers on the High Plains were too dumb or incompetent to be allowed to continue farming the region. Their feelings stemmed partly from a near-automatic reaction to the pronouncements of out-of-region experts who had never farmed or ranched in the High Plains. It also drew from a cultural disinclination to consider that there were limits to agricultural expansion and a proud reluctance to give up.[34]

This difference of outlook also came from the lack of experience in Washington, D.C. As R. Douglas Hurt is careful to point out in his history of twentieth-century farming, no one had ever tried to fix so many things all at once. The Resettlement Administration worked to move people off small farms and bad land while the Farm Security Administration labored to keep people *on* the land. The Department of Agriculture at first wanted people to leave what the secretary of labor called an "overpopulated" area, but other parts of the AAA focused on keeping folks in place and off the relief rolls. And what national planners such as Rexford Tugwell or Henry Wallace had in mind could be and often was quite different from the actualities out in the field, where local extension agents and Farm Bureau members worked to put USDA plans into effect. Much of the New Deal shared the same process: throw everything at the problem and see what works. Panhandle conservationists willingly took advantage of what they could and protested—loudly—what they disliked. But even among those Panhandle conservationists who agreed on the need for conservation as they saw it (i.e. preservation of their way of life with some adjustments for the drought,) disagreements over how to adjust and who should benefit from the new projects threatened to undo local efforts much as interdepartmental disagreements worked against the success and streamlining of New Deal programs.[35]

Problems began within the PWCA as soon as money arrived. In February 1938 Carl Hinton wrote to McCarty stating that he would resign because of the complaints that too many Amarilloans were in the PWCA's executive leadership. As Hinton put it, " [s]ince coming to Washington this time I have had some serious threats made to disrupt or break up entirely the program of the Conservation Authority, those threats being made solely because of the feeling on the part of the citizens of the Panhandle that Amarillo was unduly and selfishly interested in the same programs." Other association members dissuaded Hinton from leaving, but come May 1938 McCarty fired off his own unhappy letters to New Mexico's PWCA members about their conflicting goals. In a letter to Victor Steiner of Gallegos, New Mexico, on the edge of the High Plains, McCarty complained of a "double-cross" by Congressman Jack Dempsey, who had killed a funding measure proposed by Marvin Jones. "The thing that makes it all worse is that it involves a neighbor of ours," McCarty fumed. Fred Craddock, another New Mexican who received the same letter as Steiner, fired back on May 9. There was no "double-cross." Jack J. Dempsey had been faced with a total Department of Agriculture appropriation of $500,000 for *all* Soil Conservation Service projects, and Jones offered only an extra $15,000 for dam materials for New Mexico. Representative Marvin Jones's lack of support for New Mexico's interests led to Rep. Dempsey blocking the legislation until New Mexico got at least one dam. There was nothing personal, and he and Jones were still friends. The stress pulled at the PWCA, but the members still met to celebrate their successes.[36]

The PWCA held a meeting in Amarillo on September 8, 1938, to honor all those senators and representatives who had secured funding and support for the soil conservation projects getting underway in the Dust Bowl, and to hear what had been accomplished since the previous December. The Conservation Association members and their guests shared a jovial mood, and laughter filled the room when they learned that some delegations were unable to attend because of muddy roads and washouts. As John McCarty explained, as a result of the PWCA's efforts in Washington, "out of the whole thing there has evolved a program over and above [what the] Soil Conservation Service, various forms of Resettlement Administration and others provided for this region." Thus far $40 million had been pledged to the WPA for construction of lakes and dams and for land to be bought under the Bankhead-Jones programs, a phenomenal sum and one that was likely to get trimmed. Carl Hinton told the audience that most of the federal legislators had been unable to attend but had sent telegrams of thanks, including "Senator Hatch [of New Mexico], who I understand is stuck in the mud between here and Clovis. I suggest that this be changed into a flood control meeting." One of those who did make the meeting, Rep. Sam Guyer of Kansas City, Missouri, the ranking Republican member of the House Judiciary Committee, explained that this was his first visit to the area and that he liked what he saw. He also approved of how the people kept politics out of their regional needs. "History has proven that wherever a soil will support people, people will come until that capacity is saturated," and those same people must protect the soil and water in order to make (and keep) the land farmable, Guyer declared.[37]

Continuing on that theme, Representative Robert Hill from Oklahoma's Fifth District agreed that "in unity there is strength," something he worked for as a member of the House Committee on Agriculture. On the question of why there had been a dust bowl, he said, "God expects you to work out your problems. You have been given the right to make a living and the Congress of this country is beginning to recognize it" and to provide help for taking care of both. Representatives of several federal agencies echoed the legislators' statements, especially their praise for cooperation.[38]

The federal speakers' words seemed to suggest to some listeners that maybe the time was drawing nigh to renew calls for a major project, such as a dam on the Canadian River. A Mr. Foster of the Farm Security Board reminded listeners that his agency was "primarily interested in keeping water where it falls" and in "the principle that water should be maintained in storage in these areas." He continued, "we want you to turn in to the WPA district office a tentative proposal that will be made up of such information as the location of the dam and other information you may have readily available" and to get it on the record now. Captain Hans Kramer, visiting from the Conchas Project in New Mexico, reminded everyone that the Army Corps was already organized and ready to work on dams, harbors, and ports. New Mexico state engineer Tom McClure assured the Texans that "[y]ou may feel assured that we will cooperate in every way we can with this group." A. A. Meredith, the regional WPA deputy director, spoke to the charge of the WPA in providing the rehabilitation of people with work for "the destitute, unemployed" and said that preserving the land was part of this charge. If possible, Meredith continued, all WPA labor would be pledged to conservation efforts, as per local request. The eight hundred people attending the convivial meeting were pleased with the results and looked forward to more federal funds and projects for their own counties.[39]

Six major projects and a number of smaller ones came out of the PWCA's work. The government authorized first the dam on Tierra Blanca Creek that the Canyon Chamber of Commerce had so desired. Close behind, and completed first, came Lake Marvin (also known as Boggy Creek Lake) in Hemphill County near the town of Canadian in the eastern Panhandle. Named for Representative Marvin Jones, the lake would eventually become part of a larger national grassland and wildlife refuge. Federal and local officials christened the Tierra Blanca Creek site "Buffalo Lake" in part as a tribute to the mascot of West Texas State University (now West Texas A&M University). Tule Lake in Swisher County, Wolf Lake in Ochiltree County, Lake McClellan in Gray County south of Pampa, and John McCarty's personal dream on Rita Blanca Creek, Rita Blanca Lake, finished the list. The Soil Conservation Service also used WPA labor to construct dozens if not hundreds of check dams, small dams on arroyos and washes, on ranches across the High Plains. Hugh H. Bennett, now chief of the Soil Conservation Service, spoke at the dedication of Buffalo Lake in June of 1938, reminding everyone that the point of this lake and others was not for recreation and flood control primarily. It was instead to

> bring about a better use of the land—a type of harmony with natural conditions—a type of use calculated to ensure stability and permanence. . . . Out here in the plains country—where water is precious, where permanent natural lakes are practically unknown, an undertaking of this kind has a special significance and value. It helps to fill a long-felt need . . . to make life in this region richer and more pleasant.

But even as Bennett and area newspapers praised McCarty, Hinton, Gene Howe, and the PWCA, disagreement, complaints, and divisions within the group once again threatened to derail the hopes of the conservationists.[40]

The conservationists' disagreements were part of an area-wide discussion about urban versus rural influence in regional development. PWCA members from smaller towns expressed concern about Amarillo's role and goals in water conservation. Many people knew of the city's earlier water woes, and in 1933 Mayor Ross Rogers made the city's desire for water from Conchas quite clear. Regional critics wondered if Amarillo might be desperate enough to grab water (and economic power) at the expense of those around it. Carl Hinton remained the vice president of the Amarillo Chamber of Commerce, drawing criticism that although technically the city had only one vote just like all other members of the PWCA, Amarillo exercised more power via its overrepresentation. Amarillo also seemed to be on the verge of running out of water again; apparently a lake and well fields in three counties (Potter, Randall, and Carson) could not supply the city residents' thirst. A few Panhandle citizens grumbled quietly that yes, Amarillo was the marketing center and it did have the largest population, but there was no need to be greedy for federal money, especially since Amarillo obtained funds through both the Public Works Administration and the Works Progress Administration. The approval of Buffalo Lake on Tierra Blanca creek seemed to some, including the editor of the *Hereford Brand*, to show favoritism. Clyde W. Warwick, editor of the *Canyon Daily News*, fired back that Canyon, not Amarillo, had started the project, that Hereford was just as close to the lake as

Amarillo was, and that McCarty, Hinton, and others deserved credit instead of blame.[41]

The fissures in the unified façade of the PWCA grew deeper and more evident as 1938 passed and turned into 1939. New Mexicans and Oklahomans wanted their projects, as did other Texas counties. Wheeler County in the south presented a list of ten projects, not counting ranch tanks and check dams, while Lipscomb in the far eastern Panhandle desired sixteen new water features. Hansford County settled for a more modest seven conservation projects and Hutchinson lobbied for eight, presumably in the Breaks. As soon as tangible funds became involved, suspicions increased that someone else benefited more than they should have. Texans eyed Conchas Dam and hinted that it was their turn. Or, as the Bureau of Reclamation would later suggest, perhaps Texas could just "borrow" some of the water from Conchas to irrigate with. Some minor scandals occurred within the Conservation Association, including a guest at a meeting turning around and buying up the land that had been suggested as a possible lake site and greatly irking the locals. In another case, a local politician lobbied McCarty for even the suggestion that the county might be getting a dam in order to improve the incumbent's reelection chances. The federal agencies, notably the WPA, began cutting back budgets and projects, further aggravating area conservationists who felt their particular sections had not received their "share." However, the regional mood brightened because the rain had returned and commodity prices increased.[42]

Although four good-sized lakes and a hundred-or-so check dams and stock tanks were all that the water-stopping aspect of the Soil Conservation Service brought to the Panhandle, the New Deal agencies left a much greater legacy in the High Plains region. Civilian Conservation Corps workers had built parks and laboriously hand-cut a road and hiking trails into Palo Duro Canyon, making the colorful valley of the Prairie Dog Town Fork of the Red River more easily accessible. John McCarty had argued for this as well, encouraging people to see the canyon and to consider advertising and developing the tourism potentials in what not long before had been part of Charles Goodnight's JA Ranch. Where dirt had blown down to hardpan in Dallam County, Texas, Baca County, Colorado, and Union County, New Mexico, native grasses began returning on land once more owned by Uncle Sam. These Soil Conservation Service grasslands became the Rita Blanca National Grassland and the Comanche National Grassland, managed by the Forest Service. Contour plowing, following the lay of the land rather than cutting up and down the swells, chisel plowing, and the end of dust mulching showed some of the changes in how dry-land farmers tilled and planted, and land-grant university cooperative-extension agents, SCS employees, and others spread this information far and wide. The WPA also left colorful murals depicting local history that graced some post offices and courthouses, along with small dams, better rural roads, and other infrastructure projects.[43]

Another subtle but very critical legacy of federal activities in the High Plains came via a new network of connections. Albert Sidney Stinnett could be considered the first generation of water conservation in the Canadian Valley. He wanted to stop the waters and hold them for future use in irrigation and industry and possibly for municipal use at some point. In 1926 Stinnett had spoken a length with John McCarty, describing the possibilities of a dam on the Canadian and how the tamed waters could help the region. In turn, McCarty began making small dams possible through the Panhandle Water Conservation Association. Labor on many of those projects and for the dam at Conchas came through the WPA. The local WPA administrator was a man named Alson Asa Meredith. Meredith came to the area as an employee of Gulf Oil, was a member of the Amarillo Rotary Club for a number of years before moving to Plainview, then returned to Amarillo at the request of the Rotary to help direct their relief program in the early days of the Depression. Meredith, through the WPA, came in contact both with the Army Corps of Engineers' Captain Hans Kramer, the supervisor of the Conchas project, and with the local head of the Bureau of Reclamation while the Bureau was outlining the canal and irrigation plan that would become the Arch Hurley Conservation District. Also via the Conchas project, Meredith encountered Arch Hurley of Tucumcari, one of the movers-and-shakers in eastern New Mexico and a representative (with A. S. Stinnett) to the Canadian River Commission and the Arkansas River Basin conservation group. Meredith also knew Eugene Worley, the U.S. representative from Pampa, and Representative George Mahon of Lubbock. Government could be a very small world indeed, and Meredith and McCarty would find a way to put these connections and links to use.[44]

Although the drought officially ended in 1939, no one told the Canadian River watershed's residents. A most welcome 21.01 inches (533.4 mm) of rain had blessed the Amarillo weather station in 1939, keeping down the dust and raising spirits and crops despite the war clouds that developed over Europe late in that year. But the next year felt all too familiar as a mere 13.62 inches (345.95 mm) of moisture accumulated in the official rain gage. This was less precipitation than had fallen in a year since 1934.

But many farmers, especially south of Amarillo, were not as concerned as they had once been because now they could go outside, start a motor and bring up the underground rain. New technologies in pump design, smaller and faster motors, and better ways of connecting that motor to the pump made irrigation much more efficient, while the drought had shown the economic virtue of pumping Ogallala water. Irrigated cotton also brought in greater profits than the earlier irrigated alfalfa ever had. Ranchers, however, looked to the sky and wondered if their new stock tanks were going to be enough. The revival of the oil and petrochemical industry no doubt helped ease some of their cash concerns for those fortunate enough to receive royalty checks. And then the rains returned.[45]

People long remembered 1941. Farmers harvested a very good wheat crop in those places where it had not drowned. It rained, and rained, and rained, in Texas and New Mexico both. On the Alamositas Division of the Matador Ranch in Oldham County, the first moisture both pleased and irritated division manager John Stevens. It freshened the grass and refilled the tanks, but his car got stuck several times at the Rita Blanca Creek crossing. And the Canadian River, which divided the division's north and south pastures, came up and up. First the ranch's little coupe could no longer ford the stream and had to be towed across, a not uncommon occurrence. Then the pickup could no longer make it across. As the river swelled, it became too fast and deep even for the horse-drawn wagon, meaning that it was also too dangerous to swim the cattle and horses from one pasture to another. Roundup would have to wait, and Stevens and others were forced to visit the south pastures and the Oldham County seat at Vega by going first to Amarillo, the closest bridge, and then driving forty miles west to the ranch.[46]

Eventually the waters receded after doing the usual amount of damage. Stevens turned his attention to the screwworms and mosquitoes, made worse by the continued wet. Down in Lubbock, the usual places flooded as the city enjoyed what would prove to be the wettest year ever recorded while farmers switched from complaining about the drought to bemoaning the excess water. And in Amarillo, city leaders groaned as they read the final report from the Army Corps of Engineers concerning a dam on the Texas stretch of the Canadian. The report said that while technically feasible, the dam would not be economically sound. The dam needed to protect at least two dollars of property for every dollar spent, and there was just not enough flood risk to justify building the dam. Those who had seen the river in April, May, and June must have looked at each other and wondered if the Army Corps had studied the correct river. The Panhandle Water Conservation Association promptly filed an exemption and challenge as Representative Eugene Worley reminded *Daily News* readers that "[t]he Amarillo Project would hold back flood water of 65% of the drainage area of the stream." John McCarty thanked Senator Josh Lee and Representative Mike Morony of Oklahoma for also filing challenges to the Army Corps's report, saying, "we never dreamed of such vigorous action and cooperation." The Canadian also weighed in on the argument, or rather rushed in.[47]

Rain returned in September, sending the Pecos and Canadian out of their banks in New Mexico. The river rose in Texas as well, delaying fall roundup on the Matador Ranch, washing out fords, endangering the bridges at Amarillo and Plemons, and ruining good bottomland in the eastern Panhandle and Oklahoma. The Army Corps of Engineers announced that it would reconsider its earlier findings. Although there are no records to prove it, John McCarty, A. A. Meredith, and the mayors of Amarillo and Lubbock must have growled in irritation at the sight of all that water rushing unchecked into the sea. After all, Conchas had filled in months, not the years the engineers had predicted, and now even more water thundered downstream unused.[48]

All civil engineering–related plans and studies by the Army Corps came to an abrupt halt on December 8, 1941. The subsequent years would bring new prosperity and the resurgence of old problems, intensified by the war effort, to the High Plains and South Plains. The Dust Bowl and Depression were past, but new water woes would push Amarillo, Lubbock, and other towns to turn even thirstier attention to the Canadian.

Chapter 4

SHORTAGES AND NEGOTIATIONS

Perhaps water detests confidence. Every time Amarillo's municipal leaders announced that they had secured enough water to serve the city for a hundred years, the city promptly ran short. It happened in the 1920s, again in the 1940s, and the pattern would repeat in the future with regularity.

Everyone agreed that Amarillo and Lubbock needed water. But during the 1940s the source of that water became a bone of contention between towns, urban residents, and farmers. Lubbock and Amarillo's water quest inspired lawsuits, advertisements, negotiations, and accusations of theft of the still unbuilt dam. Irrigation, the lifesaving boom for (some) farmers on the High Plains and South Plains, was proving to be the bane of thirsty cities. Texas towns were not alone in this problem, and they shared the woes of cities around the country as postwar financial austerity collided with a national building boom. Municipal budgets stretched thin in the face of pent-up demand for repairs and new projects. Amarillo wanted a dam on the Canadian, while Hereford begged for a bridge across the Canadian in Oldham County. Borger demanded a dam and a bridge, too. Lubbock wanted Canadian River water because farmers around the "Hub City of the South Plains" wanted to irrigate with groundwater. Smaller towns on weak parts of the Ogallala aquifer competed for water with the reborn petroleum industry. And everyone wanted someone else to help pay for it all.

Urban growth and the spread of irrigation during the 1940s put the cities of the Llano Estacado on a waterborne collision course with agricultural interests and petrochemical developers. The new limits farmers demanded on municipal groundwater use pushed eleven cities to work together to obtain a dam on the Canadian. Both Amarillo and Lubbock experienced water shortages during and just after WWII, and their attempts to obtain more groundwater from nearby counties led to conflicts with the farmers and ranchers upon whom the cities' economies depended. Old ideas returned to the fore, and A. A. Meredith, along with John McCarty and Carl Hinton, began working for a dam. They turned to the Bureau of Reclamation and Representatives Gene Worley, George Mahon, and Sam Rayburn, and Senators Lyndon B. Johnson and Tom Connally to obtain the dam and a 325-mile-long aqueduct as well. But as with any major water-related venture in the West, controversy surrounded the plan even as the residents of the Llano Estacado united to obtain the largest project thus far attempted by the Bureau of Reclamation.

· · ·

The untapped floods of 1941 must have weighed on the minds of Amarillo's and Borger's city leaders as 1942 progressed and the cities struggled to find enough water to supply their surging populations and expanding economies. While they courted the Army Corps of Engineers, urging them to reconsider a dam on the Canadian, Amarillo's leaders also approached the Department of War about putting a military installation in the city. In late 1940 the City of Lubbock spent fifty thousand dollars to buy 1,400 acres (567 ha) of ranch land roughly 10 miles west of town and offered it to the Army Air Corps for a dollar-a-year lease The Army Air Corps training command accepted the offer, creating a new air base on April 15, 1941, and the first soldiers arrived in late December. Smaller towns also stood in line for air bases, and Amarillo saw no reason why it should not have one too, preferably on the site of English Field municipal airport east of the city on Highway 60. The Department of War had already leased 28,000 acres of land west of the city for a bombing range, and it seemed only logical to have an airbase to go with it.[1]

War industries swelled Amarillo's population even before official news about the air base reached the city. Ammunition manufacture required ammonia derived from natural gas. Petroleum and natural gas yielded other chemicals as well, including sulfur; since Amarillo had both gas and oil, the Pantex munitions plant opened in 1942 on 16,000 acres located just north of English Field, drawing war workers to the city. The refineries and plants in Industrial City, east of downtown, also demanded both workers and water. The Texaco refinery in Industrial City had been supplying other industrial users from its own wells, but in January 1941 had let them know that the refinery needed to cut off that supply. The city stepped in with some water but would have to find more water very soon. By April 1, 1942, when the Army Air Corps announced that Colonel

Edward C. Black would be the first base commander of Amarillo Army Air Field, almost 150 families had been moving to Amarillo each month since January, taking jobs in the new businesses or looking for places to stay while the men went to war. Because part of the city's agreement with the War Department had been that Amarillo would provide the potable water for the base, that requirement, when combined with the surging population, placed great strains on the city's utilities and on the city's new manager, Austin P. Hancock.[2]

Hancock had wells and a sewage treatment plant. What he needed were pipes and pumps and laborers to install them, all of which were severely limited by the demands of the war effort. And people flushed, brushed, washed, and irrigated their yards no matter if there was a war on or whether the water system was at capacity or had reserves. Meanwhile, demand for housing led to complaints about rent gouging, about unrealistic limits on tenants (many landlords did not allow children), and about the lack of affordable homes with decent sanitation. In May 1942 Hancock received word that the Department of War was reconsidering its original plan to use city water and instead contemplated developing a separate system for the new base now under construction. That same month the Federal Works Administration terminated funding to all non-"indispensible" projects, including the new $1,261,680 waterworks to be located near Amarillo.[3]

It is easy to imagine Austin Hancock sitting in his office that May, looking at two letters and wondering if anyone in the federal government ever spoke to other federal agencies or understood municipal sanitation. One letter came from Julius A. Krug, the head of the War Resources Board, the organization that controlled rationing for municipalities and large industries. Krug ordered that no new plumbing connections be made with pipe larger than 12 inches (31 cm) or longer than 60 feet (18.3 m), in order to curb "extravagant uses of materials for extensions of housing projects." The other note advised the city manager that forty railcar loads of 20-inch pipe were sitting in the rail yard and asked what he wanted to do with them. At least the Army Corps of Engineers had promised to take the pipe if the Army Air Force did not want it, freeing Amarillo of the expense if it came to that.[4]

Farther upstream on the Canadian, James Stevens of the Matador Alamositas Division had his own war-induced headaches to deal with. Like other ranchers in the West, he faced the problems of rationing and labor shortages in addition to the customary vagaries of the livestock industry. The "easy money" offered during the Depression by the CCC,

WPA, PWA, and other federal programs and agencies had reduced the labor pool available for the Matador, or rather it had reduced the pool of experienced people willing to labor for ranch wages. The draft exacerbated the problem to such a degree that Stevens turned to hiring boys from neighboring Boys Ranch and even postponing roundup in order to try to get the necessary help. At the same time, rationing meant that it was harder to get fuel, tires, and spare parts for ranch vehicles, while veterinarians and federal livestock inspectors faced similar restrictions on their movements. Shipping cattle became more of a challenge since train schedules shifted in order to meet the needs of the military, and as a result cattle cars had less priority than before. However, on the positive side of things, prices for cattle were higher than in some time and the rains had returned.[5]

A hundred or so river miles (160 km) downstream of the Alamositas, the war-driven demand for oil and synthetic rubber and other products caused Borger's and Pampa's populations to surge. The addition of a new synthetic rubber plant brought several thousand people to the town and created the enclave of Buna Vista, named for the Buna-N manufacturing process used at the plant. Borger still obtained its water from Phillips Petroleum, something that new city manager A. A. Meredith probably worried a little about when he had thoughts to spare from trying to collect fifteen years of unpaid property taxes from everyone in town. After all, Phillips Petroleum had gone to great trouble to find water, and as its operations expanded so did the company's own need. It would be very hard to attract new businesses, especially other oil companies, if Phillips continued to control the water. And the company's wells could go dry, unlike the Canadian River just north of Borger. An improved bridge over the river at Borger or nearby Sanford would also help Borger and Hutchinson County immensely. Somehow, people managed to adapt, working through and sometimes around wartime limits in order to get the necessary infrastructure.[6]

Using a mixture of begging, creative mechanics, and perspiration, Amarillo's utilities department managed to meet the needs of both the air base and the city. A deputation went to Washington, D.C., in February 1942 to plead with the Federal Works Agency's Defense Public Works Division for assistance in building the new airfield pipeline and also for assistance in preparing a long-term plan for water and infrastructure. Shortly after A. P. Hancock received the May letter, the Army Air Corps made up its mind, as it were, and gave the go-ahead for laying the pipe. City workers and contractors put down 51,000 feet (15,555 m) of heavy cast-iron water mains from the city out to the new airfield in two weeks during July, including laying one mile (1.6 km) on one day. At the same time, Hancock scrounged enough materials to connect the newest wells in the Palo Duro well field south and west of town in Randall County to the city's water system, augmenting supply enough to fend off shortages, although he did ask people to quit watering their lawns quite so heavily. He must have bitten his tongue a little at the city commission's optimistic June 26 pronouncement that it was going to sell six sections (1,554 ha) of land that it owned near the Randall County well field because "the Commission thinks that it has ample water on other lands to supply the city for its present and future needs." At the turn of the year the Federal Works Agency granted the city $422,052 for construction of the long-delayed waterworks.[7]

The creation of sewer connections proceeded more slowly and did not meet prewar construction standards. Outflow pipes of only 18 inches (41 cm) in diameter ran a foot or so below ground, if that much, and in at least one place (West 3rd and Hayden Streets in north Amarillo) metal plates covered the pipes to keep them from rising to the surface as the ground swelled or shrank with the seasons and moisture. Nonessential projects accumulated, and the city addressed them as time and materials permitted, leading Hancock to remind the commission that "[t]he present [sewer] line was built years ago and was to accommodate a town of 40,000 people and it was now running full capacity," and they needed to keep this in mind should a new industry move into the city. This was true in many other cities in the United States, and one imagines Austin Hancock commiserating with A. A. Meredith and other city managers.[8]

Lubbock, to the south, faced similar problems. Although the city did not supply water to the air base, Lubbock's water supplies suffered growing pains nonetheless. In April 1940 the city obtained a $10,000 short-term cash loan for water, sewer, and street extensions because of "a rapid growth of population." Residents with their own wells complained that nearby city wells "dried up" the older private water sources. As happened to Amarillo, in September 1942 the War Production Board informed Lubbock that the city could no longer extend water service lines to new constructions, forcing the utility department to come up with new ways to distribute (and meter) water. Things stabilized somewhat by late 1943 but in July 1944 the city planning commission recommended that Lubbock hire an engineer to look into water supplies enough for a city of 100,000 to 150,000 people. "The report and investigations [are] to include both underground and surface supplies."[9]

Why surface supplies? The city's leaders realized very quickly that although they could acquire water for Lubbock from outside the city, the municipality faced considerable resistance when it did so. Irrigated cotton had become a major industry and one that the City of Lubbock hesitated to compete with. Under Texas laws, nothing could stop the city from buying water rights, drilling wells, and pumping as much water as was needed, so long as the city obtained the necessary rights-of-way and materials. However, farmers took a dim view—and complained loudly— of the city "stealing" their water and threatened to take their business elsewhere. Obtaining materials for such pipelines also posed a major problem, one that Lubbock's leaders hesitated to take on except where absolutely necessary.[10]

South of Lubbock, smaller towns faced water shortages caused by geology and exacerbated by petroleum exploration. The water-bearing Ogallala aquifer grew thinner the farther south on the Llano Estacado that one went, eventually tapering out of existence south of Lamesa. The thinner sands and gravels held less water, so wells drilled in the formation failed sooner and yielded less when they did flow. Tahoka-area water users faced the additional challenge posed by contaminated groundwater. The water table reached several natural salt lakes that leaked brine into the Ogallala during some parts of the year, making the water downstream of the lakes unusable for domestic consumption or irrigation. During the 1930s Tahoka and Brownfield experienced water crises when their wells failed and new wells proved useless. Rain in the late 1930s and early 1940s helped the situation, but during that same time the arrival of oil exploration rigs added a new, very large, water consumer to the mix.[11]

Well drilling required even more water than it had just a decade and a half before. The old cable-tool drilling process described earlier used at least 42 gallons (159 l) of water per eight or ten feet of hole drilled. The "modern" rotary drilling process consumed much, much more water because it depended on a constantly circulating mixture of water, chemicals, and clay or other thickeners ("drilling mud") to carry the stone chips up out of the hole. Some of this mud could be reused, but the water still had to come from somewhere. South of the Llano Estacado, oil companies in the Permian Basin tapped a very deep, brackish aquifer and left the fresh water for other uses. In the Northern Basin Platform, the geologic name for the oil-producing area under Lubbock, Gaines, Hockley, Dawson, Terry, and Cochran Counties, drillers tapped what they could find of the Ogallala. Area residents appreciated the new incomes from royalties and from oil-field wages and oil-field businesses, but the prosperity came at a high water price and sent the leaders of Lamesa, O'Donnell, Tahoka, Brownfield, and Lubbock searching for water far from home.[12]

As the larger cities looked for sources of pipes and labor, the hunt for water sources did not completely stop. Amarillo and Lubbock drilled new wells, Lubbock within or very close to the city limits and Amarillo in the Randall County well field. Amarillo incorporated two new subdivisions and bought their wells and pipes for the city. Carl Hinton of the Panhandle Water Conservation Association returned to Washington, D.C., on behalf of Amarillo and the rest of the PWCA in May 1942 to gauge the tone of the various federal agencies and to see if there might be a way to have a Canadian River dam or other smaller projects funded as part of war-production efforts. After all, the water would be helping wartime industries, or so the argument ran (and would run again later). In his letter approving the funds for Hinton, businessman W. A. Warren of Canyon inquired about getting permission from someone, probably the U.S. Department of Agriculture, to raise the Buffalo Lake dam so that "a heavy stream then could be released during the summer months and be of great value to all farmers below the dam."[13]

Although neither request was granted, representatives of the Bureau of Reclamation attended the June 30 meeting of the Amarillo Chamber of Commerce and advised the chamber members that the Bureau had a postwar water project in mind for Amarillo. The Bureau was still working on an irrigation project for the area downstream of Conchas near Tucumcari, and knew of Amarillo's interest in the Canadian. "The project has three purposes," the Reclamationists explained. "1) to supply water for irrigation for farmers along the route, 2) to supply ample water here to provide for the needs of a city many times the present size of Amarillo, and 3) to provide a tremendous lake for recreational purposes near the city." The water would come from Conchas Lake via an open canal. But as everyone in the meeting knew, bringing the vision of irrigated acres and strings of fresh-caught crappie to life would have to wait for the end of the war.[14]

Despite the creativity of Hancock and his assistant, N. V. Moss of the city utilities department, by the time WWII ended Amarillo suffered still more infrastructure problems. Rationing continued even after fighting stopped in Europe and Japan. People began wondering why the city could not do anything about the daily 8:00 a.m. sanitary sewer backups and overflows at 3rd and Hayden. "Because we told you when you built there that the sewer connections were at their maximum" formed the gist of the rather more tactfully phrased response. Complaints about sand in the pipes in Wolflin and other south Amarillo neighborhoods offered Hancock the opportunity to explain just how tight the city's water budget was: from the time it left the ground in Randall County the water never stopped moving long enough for the sand to settle out as it passed through the pumps, treatment plant, water towers, and delivery lines. Demand equaled supply with almost no surplus.[15]

On June 18, 1945, a dry summer and a pump failure terminated what little surplus existed. Amarillo pumped 16 million gallons and city residents used 17 million gallons, even though the city parks department suspended watering that morning. Rationing came into effect until the city's newest wells were tested and brought online five days later and water reserves rebuilt. As the editorial "These Times" in the *Amarillo Times* explained on April 8, 1947, Mayor Joe Jenkins's administration faced two major problems: getting adequate water and storage, and laying sufficient sewage line "to serve a growing city." Although the city had installed more than forty-one miles of pipe, fifteen wells and associated pumps, and two reservoirs, along with connecting "1,114 homes, one factory, 25 businesses, one feed mill, one swimming pool, one park, two schools, four churches and a welfare station" to the sewer system between October 1941 and April 1947, things remained tight. No one could deny it: Amarillo had run out of easy water—again.[16]

The situation grew more problematic as the summer shifted into fall, even as the city's leaders began working on solutions. In September 1946 Mayor Joe Jenkins appointed a citizens' Water Advisory Board to look at all possible water acquisition options. Members included J. B. Briscoe of the Santa Fe Railroad, two contractors, the president of the local Coca-Cola bottling plant, and an insurance salesman. In an editorial printed the same day as the board was announced, Gene Howe supported the proposed water and sewer bond issue but with reservations. "We . . . believe that the city should work constantly toward a long-range, permanent water supply with the possibility of a Canadian River project being explored thoroughly." On October 5, Secretary of the Interior Julius Krug visited Amarillo to inspect Interior's helium plant, located west of Amarillo. With him came Michael W. Straus, the head of the Bureau of Reclamation. The helium plant sat on the southern edge of the Breaks, not far in a straight line from the river, so perhaps that explains in part why Straus came along. The Bureau also had a regional office in Amarillo, a more likely cause for Straus to spend time in a place without any Bureau projects. Either way, the currently available sources are vague about the head of the Bureau's visit. Four days later heavy rain pummeled the area, causing highways to overflow and sewers to back up (again). The old eighteen- and twenty-four-inch lines could not take the rainwater and sanitary water together. That problem would be relatively easily solved once the city had resources and supplies to lay new and larger pipe. Finding more water would not be as simple.[17]

Where to go for more? Expanding the Randall County well field would become counterproductive, the city engineer and manager knew. The closer wells are spaced, the less each can draw before the surrounding water level declines enough that the "cone of depression" affects neighboring wells and vice versa. Plus the city did not want to upset farmers and other landowners more than necessary by taking too much water from Randall County. Other options lay farther southwest, in Deaf Smith County, and northeast in the city's well field in Carson County, although regional well-drilling expert D. L. McDonald had already warned against looking in that direction for water. The Pantex Munitions Plant had closed when the air base did, and Texas Tech University had bought the land but not the water, so the Pantex site remained a possible option but one the city hesitated to push too far. The city stopped delivering water to the now-closed air base, but that did not increase the total supply. After considering various options Mayor Lawrence Hagy and the city commission opted to turn southwest, to lands straddling the Randall–Deaf Smith county line. The decision made economic and engineering sense because there was proven water at relatively shallow depths and the rights of way would be shorter because only a connection with the existing pipeline from the Palo Duro well field was needed. In contrast, ongoing well-level declines in parts of Carson County made the city leery of compromising what it had in that county.[18]

Not everyone agreed with the city commission's decision. D. L. McDonald, the pioneer irrigation well driller in Deaf Smith and other counties in the area, had already voiced his objections in a series of large ads placed in Amarillo's newspapers and in a booklet that he sent to anyone who expressed interest. He firmly opposed Amarillo's plans to tap southwest Randall County's and eastern Deaf Smith County's groundwater. "Many think the water . . . best left for use by the irrigation farmers of Randall County." Furthermore, the Deaf Smith water was too shallow and limited, as was Oldham County's (due west of Amarillo). McDonald knew where deeper, more reliable water existed and had offered to tell the city but had been rebuffed. "To pass up one of the largest potential water supplies in this country on 'economic grounds' for an admittedly temporary supply of short life would indeed be a tragedy in the affairs of

Amarillo," he proclaimed in one article. McDonald claimed to have only the public's interest at heart and declined to publicly name the location of the better water in order to forestall a grab by developers or others who only wanted to profit from others' need. As T. E. Johnson of the *Amarillo Times* explained in an editorial column that May, McDonald did not want to see $2 million wasted and was staking his reputation as the region's groundwater expert. Instead, McDonald (according to Johnson) believed that the city should buy the water under the Pantex facility and he "scoff[ed]" at the idea of using surface water. There had been too much talk about the higher cost of pumping from deep wells at the Pantex site versus the "shallow water" in Deaf Smith County. Despite McDonald's protests, in the fall of 1947 the city commission and Amarillo's newspapers announced the route and general location of the new field in southeastern Deaf Smith County, and all hell broke out.[19]

Loud objections came from several individuals and groups. The farmers in Randall and Deaf Smith Counties protested vigorously against Amarillo's water grab. It was bad enough that Amarillo had driven the water conservation projects like Buffalo Lake and public works such as developing Palo Duro Canyon (a CCC project). Now the city wanted water that should go to landowners and real farmers who paid taxes and grew the crops that made the city's prosperity possible. The William Bush family, owners of much of the land through which the city's new pipeline would run and owners of irrigated farmland as well, filed an injunction blocking the purchase or condemnation because the price was too low and because they too wanted to safeguard future water development. The value of the land would decline, even as rangeland, if the water went away or the city claimed a large swath cutting through the center of the property for a pipeline. Randall County farmers waded into the fray and reporter Cal Brumley warned, "[T]hey will be as tenacious as Johnson Grass in fighting to keep the city out of the area." Emotions ran high at the January 27 city commission meeting, to put it mildly, as farmers demanded that if the city used water and interrupted irrigation, the city would have to pay for the damage. Amarillo's governing body tried to move forward but as another headline stated, "Second Suit Blocks City Water Plans" as the Bush estate countersued against the proposed condemnation. Amarillo's city commissioners decided that the possible water did not justify the pain.[20]

A tactical retreat seemed to be the order of the day, and in spring 1948 Amarillo opted to expand the Carson County field and concentrate on finding water under rangeland. Lawrence Hagy, A. A. Meredith, and others also looked north to the Canadian River. Six months earlier, the first legal steps necessary for obtaining river water had been explained when Colonel E. V. Spence of the state Board of Water Engineers visited the city. He had pointed out that Texas, New Mexico, and Oklahoma had to draw up a river compact before anything else could be done. He had also urged the city to pick a well site farther from the current well fields, and Hagy and others may have decided that heeding his advice was a good idea—this time. After some discussion in the commission, Mayor Hagy contacted Congressman Eugene Worley in early April 1948 to see how to go about obtaining federal assistance for a dam or other water-retaining or -diverting project. Hagy explained to the press that although Amarillo could not afford the dam at that exact moment and doubted "that the government would consider it at the present time, [nevertheless] a project like this should be worked up so when a public works program is instituted, this dam would be high on the list." The next logical question was: which entity should build the dam?[21]

As before the war the city found several options, one of which lay with the Army Corps of Engineers. The Army Corps had returned to the dam-building business with a vengeance, as if making up for lost time. As a result, the Corps restarted its survey of the dam-site possibilities in the Canadian River Valley in Texas. The Corps remained the logical choice for several reasons: first, they were already familiar with the river, having built Conchas and surveyed the other reaches of the stream; and second, there would be much less, if any, cost to repay from an Army Corps structure as compared to Amarillo hiring a private contractor. But as it had before, the Corps focused on navigation and flood control, not irrigation or municipal supply. Nevertheless, the Corps surveyed the valley and considered three dam sites in particular. As army engineers studied the Canadian, Amarillo's leaders looked at their second option.[22]

Given the nature of the project, the Bureau of Reclamation had become a definite possibility. In 1942 the Bureau had spoken to Amarillo's conservation-minded leaders and the Chamber of Commerce, suggesting that extending a canal from Conchas Dam to Amarillo might be possible once the war ended. This canal would have been primarily for irrigation, and this option ignored the question of whether the Canadian could supply enough water to make the hundred-plus-mile trip through a dirt-walled canal and still irrigate any economical acreage. The Bureau remained in the irrigation and reclamation business by definition, and any municipal water it provided came as a sidelight to the main irrigation projects. Yet the Hoover Commission's *Report on the Executive Branch of the Government* of 1949 would argue that reclaiming wastelands for towns could come under the Bureau's auspices, if done carefully. Congressional legislation already passed in 1948 allowed the Bureau to develop municipal water supplies with the same repayment requirements as irrigation projects. According to Mayor Hagy, he understood that "Bureau officials favor installation of a dam on the Canadian. The Bureau has a regional headquarters in Amarillo." So Hagy, A. A. Meredith, and others kept the Bureau in mind as they waited to hear from the Corps. After all, the Corps did not ask to be reimbursed, or at least not as much as the Bureau required. Meanwhile, other High Plains residents watched the survey with great interest.[23]

Amarillo was not the only municipality interested in acquiring more water. Borger depended on Phillips Petroleum and wanted other options because, as Representative George Finger later put it, "[it] handicapped the community to a great extent in that they had no water to operate industries to come to the area." Pampa, to the east in Gray County, depended on relatively deep wells and had since the town's founding in the late 1880s. The papers of the White Deer Land and Cattle Company describe the difficulties the railroad faced trying to reach water enough to refill locomotive tanks. Town developers fared a little better but still dug more than 250 feet (77 m) down to find water. Pampa was not yet short of water but certainly would benefit from having more, especially from a visible source. Well to the south of the Canadian watershed, the communities of Lamesa and Tahoka faced shortages as already described. Between Lubbock and Amarillo, Plainview possessed abundant groundwater but expressed interest in learning more about the plan if Lubbock intended to tap the Canadian.[24]

To the west of Amarillo and south of the Canadian, the town of Hereford in Deaf Smith County did not need water but still wanted a dam on the Canadian. What Hereford needed was a bridge or a road over a dam across the river. Texas State Highway 81 would then run from Dalhart as far as the Big Bend, bringing traffic, tourists, and economic development to Hereford, assuming travelers had some way to cross the Canadian that did not involve waiting for low water or diverting seventy miles to Amarillo and back. The state had bought land for the right of way, but the Texas Highway Department ran out of funds before it could build the bridge. As time passed, Hereford residents would become very vocal proponents of a dam but only if it was built in their preferred location. Hereford's interest and conditions give a hint of the tensions that were developing beneath the cities' cooperation.[25]

However, that remained in the future. Representatives from Amarillo, Borger, and other northern towns met in early May 1948 to talk about the area's needs and how to go about obtaining the project on the Canadian. A spokesman from Lubbock sent word to Amarillo that the Hub City of the Plains did not take kindly to being left out. "The omission aroused curiosity in Lubbock as to whether anyone in Amarillo proposed to try to freeze this city and area out of participation in a project which might hold the permanent answer to the water supply problem of the entire region." Mayor Hagy quickly replied that nothing of the sort had been intended and that Lubbock would be included in future gatherings if its citizens so desired. As a result, on June 29, 1948, the Panhandle Water Conservation Association organized a major meeting in Amarillo, ahead of the Army Corps of Engineers' issuing its final report. Representatives from cities and towns around the Panhandle, including Borger and Vega, as well as businessmen and conservationists like John McCarty of Amarillo, Arch Hurley from New Mexico, Gene Klein of Amarillo's Chamber of Commerce, and corporations such as Southwestern Public Service Company (SPS) discussed projects and options. Because its mandate was too broad in scope for the proposed project, the PWCA, such as it was after six years of inactivity, bowed out as the legal and organizing body. Instead, an executive committee comprised of people from Amarillo and Lubbock began gathering the economic information needed by the Army Corps and Bureau to justify building a dam on the Canadian. The group was called the Canadian River Water Users Association, later adjusted to the Canadian River Municipal Water Users Association and the Canadian River Project Organizing Committee (CRPOC).[26]

Once inspired, High Plains residents worked diligently to get what they wanted despite negative reports and minor internal dissent. Although the Panhandle Water Conservation Association had lapsed into inactivity during the war, the expertise remained available through Carl Hinton, John McCarty, and now A. A. Meredith. It also helped that Eugene Worley, the rancher turned U.S. representative from Pampa, and George Mahon, representing the South Plains and Lubbock, had gained seniority in Congress, Mahon sitting on the House Appropriations Committee. Speaker of the House Sam Rayburn was inclined to assist fellow Texans, as were senior senator Tom Connally and junior senator Lyndon B. Johnson. A. A. Meredith's contacts from his days as regional director of the WPA would also prove very useful, especially when the Army Corps' report threatened to kill any hope for a federally built reservoir on the Canadian.[27]

In mid-June 1948, the first word reached Hagy, McCarty, Meredith, and other regional leaders that the Army Corps's report would probably be negative. Technically, no problem existed with building a dam on the Canadian River in Texas, and technology existed to pump the water up to Amarillo, Borger, or wherever else it was wanted. There would be flood protection benefits for those downstream of any dam. However, those benefits did not equal the cost of the dam. The Army Corps of Engineers could build a flood-control dam only if the damage prevented was greater than the cost of the dam by a factor of at least 1.2 to 1.0 (for example $1.2 million worth of property not damaged in floods versus a dam that costs $1 million). Even if they added in benefits to wildlife, the Army Corps could not justify the expense. Working quickly, Eugene Worley, George Mahon, and Texas governor Beauford H. Jester asked the Corps to hold back presenting the final report to Congress while the Texans looked at other options, including the Bureau of Reclamation.[28]

Regional conservationists had been talking with the Bureau as a fallback since 1947, and the Bureau was already in the process of doing a full survey of the Canadian River as part of a study of the entire Arkansas River Basin. Initial reports from the Bureau were positive, and in fact the Army Corps gave the data it had collected to the Bureau to help speed the process (and possibly to deflect the Texas delegation's ire). At the same time, aware that no one could utilize the river's waters without an interstate compact, John McCarty organized another meeting of representatives from the governors of Texas, New Mexico, and Oklahoma on August 3. The states' representatives agreed that they needed engineering studies along with much more detailed information about the river and its flow before anything could be negotiated. Arch Hurley from New Mexico added what would prove to be a critical caveat, warning that the Pecos River Compact had complicated things greatly and hoping that a Canadian River Compact would be less problematic. Oklahoma just wanted its share of the water, pointing out that the state had lost too much "$300 per acre farmland" to the capricious Canadian's floods and that the Army Corps did not consider the value of lost crops in its figures. On a more local level, Amarillo and Lubbock pooled their funds and hired "a leading authority in the field of industrial engineering and business economics," Burt C. Blanton of Fort Worth, to do a full economic study of the region to show the positive effects a dam and reservoir would generate, and the negative effects of leaving the area dependent on groundwater. Over three hundred pages of tables, graphs, charts, and economic rankings seemed to prove that if the area was to continue to thrive, it needed more than just wells for water.[29]

The proponents of river development put their funds where their votes were, encouraged by the Bureau of Reclamation. In October of 1948 Amarillo's Chamber of Commerce donated up to $2,125.00 to help fund the engineering survey, with Lubbock pledging funds and as demographic and economic data. The 1,500-member-strong Panhandle Outdoor Sportsmen's Club had already pledged its support for the dam and encouraged members to assist in any way necessary. On October 7 Michael W. Straus of the Bureau met with Hagy, McCarty, and Jack Cunningham of SPS. Regional Bureau director H. E. Robbins reminded Amarillo's leaders that they had to keep the Army Corps's report on the back burner until the Bureau finished its survey. Next they had to renegotiate the Canadian River Compact. Only then could the Bureau start work. Straus had been in contact with Eugene Worley, a good sign. John McCarty asked about the possibility of the dam as a public works project a la the WPA dams and Straus said no, reminding McCarty (and the listening reporters) that "[t]he people of the Panhandle are in no mood to wait for a depression to make sure of a water supply." Come December the massive report arrived from Fort Worth and confirmed what Lubbock, Amarillo, Borger, and other towns' residents wanted to know—in Blanton's view, using demographic data from the University of Texas and economic information from the High Plains, a reservoir on the Canadian was absolutely necessary for the continued growth and success of the High Plains economic region. It was a Bureau-worthy project, a vital necessity for a region so recently forced to lean on the federal government for support, and now the cities had the data to prove it.[30]

While the Bureau worked, so did High Plains dam promoters. Shortly after the meeting in October, Amarillo's city commission passed a resolution strongly in favor of a dam and pipeline. In November 1948 the Bureau had argued that the next four years "present the best prospects Amarillo has had for obtaining the dam," provided that certain things got done. The creation of the CRMWUA was one of these, in that it allowed the area to participate in the Texas Water Conservation Association programs, putting the Canadian into the same framework as other state rivers in matters of water conservation. Lawrence Hagy went to Washington in February 1949 to meet with Julius Krug of the Department of Interior, Representatives Worley and Mahon, and Senators Lyndon Johnson and Tom Connally to keep them informed of the progress thus far. Federal Judge Marvin Jones (a former U.S. representative), Lubbock's water representative Clarence Whiteside, and Howard Robbins of the Bureau also attended some of the meetings. Two weeks later a joint meeting of the Bureau, Lubbock, Amarillo, and the PWCA was held to plan for a region-wide gathering in Plainview later that month. Plainview sat roughly halfway between Lubbock and Amarillo, and the distance allowed John McCarty to make a point about the Bureau's skills. "After all, if they could build the Friant-Kern canal in California we can do the same thing here." Five days later fifty-six people, including representatives from Amarillo, Lubbock, Plainview, Borger, Slaton, O'Donnell, Pampa, Levelland, Dimmit, Tahoka, Post, Floydada, and Littlefield met with engineers from the Bureau of Reclamation to make formal plans.[31]

The meeting covered a broad range of topics that returned again and again in the future. Harry Burleigh, an engineer from the Bureau, discussed the data that had been collected to that point and tentatively suggested that Tascosa was the most promising dam site. "The Canadian at this point has a firm water potential of 100,000 acre feet annually," he advised, adding the warning that the water would not be as high quality as well water. Mayor W. A. Rogers of Lubbock replied that his city did not care—they needed the water. As he had said in August of the previous year, the city had wells running dry. Irrigation farmer and Lubbock resident Hobart Nelson agreed and said he would pay the higher taxes "that it would require to junk Lubbock's entire local water system if so doing would add to the underground water potential for irrigation." A. A. Meredith of Borger stated concisely that Borger had a water problem and needed more water: one business alone used 16.5 million gallons *per day*. R. O. Stark, the manager of the small community of Tahoka south of Lubbock, also had no problems with the higher cost of imported water, since Tahoka already paid twenty-five cents per thousand gallons of water, some of the highest rates in the region. Only the city of Plainview faced no water woes, but their representative indicated that the city remained interested in adding to the city's supply if things progressed that far. Ten counties had towns that wanted Canadian River water, which might be enough to raise the funds needed to pay for the project, however much that would be. The Canadian River project was now in the Bureau's lap, so to speak.[32]

The Bureau of Reclamation produced its report in June 1949 and said almost all that the High Plains conservationists could have wanted, even as the price tag took their breath away. Yes, a dam was quite feasible, and there were three possible sites available. Although the dam would provide limited flood-control benefits, given the rarity of major floods since the closure of Conchas Dam, it would provide enormous municipal water supplies. The regional economy was strong enough to support paying off the dam and aqueduct in the usual fifty-year time span. The most contentious portion of the report proved to be the dam's location. Contrary to earlier statements, now the Bureau favored a site near the hamlet of Sanford, Texas, upstream of Borger. A dam near Amarillo required too much movement of infrastructure, and the local landowners, major financial presences in the area, were not entirely pleased with the idea of losing prime rangeland and water. Tascosa, the westernmost option, had acceptable geology, but it would require a larger dam to hold back the same volume of water as would a dam at Sanford, it would catch less water than would downstream structures, and the closest source of the rocks necessary for the proper riprap lay in northeastern New Mexico; also the pipeline would have to run farther, costing more. The final bill before overruns or inflation came to $85,383,000, of which $77,892,000 would have to be repaid by the cities over fifty years.[33]

Even as they worked together and savored the welcome, if eye-wateringly expensive, good news, differences in motivation and methodology remained among members of the CRMWUA. One of the disagreements that arose early centered on advertising the Canadian River Project's purpose. Amarillo's and Borger's business leaders and newspapermen talked about supplying drinking water and securing a visible water supply for industries. Groundwater remained mysterious and uncertain, but a large surface reservoir would reassure any factory or corporation interested in moving to Amarillo, Pampa, or Borger. Despite the "little conflict" with Deaf Smith County's farmers, agriculture did not play a role in Amarillo's justifications for a lake on the Canadian until late in the lobbying process. Lubbock, however, focused on agricultural concerns, and at one point Clarence Whiteside of the Lubbock Water Development Board scolded his counterpart in Amarillo over the matter. Everyone in the country wanted industries to move to their city, Whiteside argued, and talking about it might well prejudice outsiders against the project and eliminate vital federal support. Instead, Lubbock advertised, "A gallon of water from the Canadian is a gallon saved for irrigated agriculture." The motto appeared in letters and on stationary as well as in pamphlets and news stories about the efforts to obtain a dam on the Canadian. More oriented to ranching, the oil business, dry-land wheat, and transshipment, Amarillo's business community acknowledged Lubbock's protestations but did not really change its lobbying focus. The differences between the cities would become more evident as time progressed, but in 1949 the eleven interested towns presented a united front—more or less.[34]

Instead, what opposition there was came from landowners in the

Canadian Valley and proponents of the Tascosa site. Mrs. Sanford, who owned the land in the valley that would be under the easternmost proposed dam site, was not enthusiastic. "It would ruin a good cattle ranch" that she had owned for over fifty years. The John Fain family, owners of some of the property around the Tascosa site, had similar objections: a dam was fine in itself, but not on their land. One wonders if they had doubts about the possibility of being paid what they felt the prime rangeland was worth. Another opponent, and probably the most vocal in his objections, was J. Evetts Haley. His family owned land on the river in Hutchinson and Roberts Counties that would be affected by any dam. However, his difference stemmed more from philosophical disagreements rather than financial. These derived from his experiences during the 1930s with the AAA and other federal agencies. He remained suspicious of all federal projects, especially those where the federal government remained involved after the construction work finished. Haley also ran against Representative Gene Worley for U.S. representative in 1948, but it is unlikely that his election campaign affected his views on limited government. Aside from the rancher-historian, opposition to the project focused on those whose lands would be directly affected and those concerned about taking on too much debt. There were no complaints lodged on what would later be called environmental grounds. Conservation of water meant wise use of it, and catching the destructive floods seemed to be a very wise and beneficial use of the otherwise wasted water. The dam backers only needed to persuade the federal government of that salient fact, and to soothe all the ruffled feathers south of Tascosa, and all would be well.[35]

The rumors of the Bureau's choice of dam site generated the most controversy over the project prior to the state board of water engineers permit hearing in 1954. George Mahon must have shaken his head a little in May 1949 when he began receiving missives from the Levelland Chamber of Commerce, constituents, and the Hereford Chamber of Commerce urging that the Tascosa dam site be selected. People from Littlefield especially felt betrayed by the Sanford announcement because as early as November 1948 John McCarty had told them that Tascosa was the most likely site. The arguments against Sanford and Amarillo focused on hydrology and finances and grew rather more aggressive as the months passed. A dam at Tascosa would let water percolate into the underflow and recharge the aquifer even as far as Plainview. It would be cheaper to pump the water from Tascosa to Amarillo and Borger, since it was downhill (at least, it was once pumps lifted the water out of the Breaks). The project would be spared the costs of relocating Amarillo's sewage treatment plant, two gas pipelines, and numerous oil and gas wells. And State Highway 51 could be run over the top of the dam, bringing benefits as far as the Big Bend and easing traffic on U.S. Highway 87/287. Pete Cowent of Hereford declared that "[w]e're sure as hell not going to quit." Cowent also drew on the late D. L. McDonald's argument that the Canadian recharged the Ogallala to back Hereford's and Levelland's claims to a dam at Tascosa. The oil and gas business would last at best fifty years, but farmers would be around for two hundred at least, and so they deserved the underground water provided by a dam at Tascosa. All this occurred before the Bureau even released its initial report or anyone proposed legislation asking Congress for any dam.[36]

Armed with a positive report from the Bureau, pounds of economic data, and very heavy political clout, Gene Worley called for hearings before the Rivers and Harbors Committee on July 11, 1949, a few weeks after Congress received the Bureau's final report. Representatives from Amarillo, Lubbock, Borger, Pampa, Plainview, and other towns spoke eloquently about the need for and benefits from the dam. It would add certainty by providing a guaranteed and visible water supply. The water would spare vital agricultural irrigation supplies, thus assisting American food security and enhancing regional and national self-sufficiency. The river water had less fluoride than did Ogallala water, so that children would no longer have brown teeth. The cities could easily repay the costs of a dam and 350-mile-long (563 km) aqueduct, and it would help prevent a repetition of the economic hardships and need for relief caused by the drought of the 1930s. Mayors, city managers, John McCarty, A. A. Meredith, Worley, and others all took turns arguing for congressional approval of a dam and aqueduct on the Canadian River, and each repeated their concern for the residents' dental health.[37]

Eugene Worley had introduced House Resolution 2733 in June 1949, and all involved hoped that it would pass easily. High Plains residents appeared so confident that T. E. Johnson of the *Amarillo Times* penned a cautious editorial about excessive optimism. The bill needed to go through a great number of legislative steps before it could be considered certain, he reminded readers. In contrast the *Amarillo Globe*, always more in tune with the city's movers and shakers, declared that inquiries "[find] dam prospects rosy." Alas for the *Globe*, several congressmen raised objections to Worley's bill. As proposed, the Canadian River Project would be both the largest project the Bureau had attempted thus far and the most expensive. Representative Walter K. Granger of Utah, no stranger to Bureau projects, informed the committee that he needed more information as to how exactly this enormous project would work and (more importantly) how the cities intended to pay it off. The system set a new precedent for size and scope. In addition, no other Bureau project thus far served only as a municipal water supply, and Representative Granger wanted all the details. After talking with Worley and a representative of the Bureau, Granger announced himself satisfied, and the bill passed out of the Rivers and Harbors Committee unanimously.[38]

The bill sailed through the full House of Representatives on August 4, 1949, raising some eyebrows in the Senate. The legislative year had drawn almost to a close, and Speaker Rayburn along with Eugene Worley and George Mahon wanted the bill out and over to the Senate as quickly as possible. In order to do this, they rushed the Canadian River Project authorization onto the House floor without discussion and voted without further consideration of the $80 million authorization bill. The proposed law traveled across the Capitol to the Senate chambers, and Worley and George Mahon let their constituents know the good news.[39]

The legislative brakes came on in full force once the proposal reached the Senate. One of Missouri's senators quickly cautioned his colleagues in both the House and Senate that a due respect for the duties of the committee were in order and that the appropriation request and authorization legislation would receive the same careful scrutiny as every other bill presented. Senator Clinton P. Anderson of New Mexico emphasized to everyone that the questionable tactics used in the House were not acceptable to the august body that was the U.S. Senate.[40]

Senator Anderson's cool response stemmed from problems with the Pecos River. Texas and New Mexico had negotiated an agreement pertaining to how much of the Pecos River's waters were guaranteed to Texas by the upstream state. Objections over water amounts raised in 1946 by Senator Carl Hatch of New Mexico had slowed negotiations and delayed ratification of the Pecos River compact for two years, so that it only came into force in 1948. Anderson informed the Texas delegation and the people of the Panhandle that unless New Mexico performed a full study of the Canadian River and the state engineer gave his approval and confirmed that any interstate agreements between Texas and New Mexico over the Canadian River were hydrologically and legally sound, New Mexico would not support additional dams on the Canadian. First, New Mexico needed a river compact, approved by the New Mexico state engineer. Then Texas could have a dam, all other things (finances notably) permitting.[41]

Citizens of Amarillo, Lubbock, and the South Plains reacted with heat and fury. How dare New Mexico obstruct a dam in another state? Had not Amarillo's citizens helped pay for Conchas? Senator Anderson's unreasonable refusal was a near betrayal of the Texan members of his fraternal organization. George Mahon, A. A. Meredith, Lawrence Hagy, and others tried to calm the rhetoric and to work out a compromise. It is surprising that New Mexico's conditions caught anyone familiar with rivers in the West off guard, but the Texans were. In early September a delegation from the Panhandle went to Santa Fe and Albuquerque to negotiate with Anderson and the governor. The Panhandle men, after several hours' discussion with Anderson, agreed to the delay and compromise. Senators Johnson and Connally agreed as well and Anderson lifted his hold, allowing the bill to leave committee. It would be spring of 1950 before the bill reached the Senate floor, but it made progress nonetheless.[42]

Opposition to the Canadian River project also came from inside Texas. Elements within the West Texas Chamber of Commerce told the *Amarillo Globe* that the Bob Baskin Dam on the Brazos River should come before the Canadian project. This brought a hasty letter from L. A. Wilkie of the Chamber assuring Plains residents that no, that was not the official position of the entire Chamber. What the Chamber wanted was a law stating that drought losses would be calculated in federal dam studies just as flood losses were. A month later a long editorial in the *Fort Worth Star-Telegram* chastised High Plains planners for leaning on the federal government. Using language much like that spoken by J. Evetts Haley, the editor wondered why West Texans were so quick to reject state management of water and to lean on the federal government for aid. Had the state not just fought the federal government all the way to the Supreme Court in order to protect Texas's right to oil from its tidal waters in the Tidelands suits? The Canadian River project and similar ones would "put the Secretary of the Interior directly into the municipal and industrial water supply business." Should not that be a state prerogative, with the payments going to the state instead of the federal government? The editor concluded that "if Texas is to retain any sort of control over its water resources, the project and the water rights should be owned, operated and maintained by the people paying the bill, not by the Secretary of the Interior."[43]

Word of the Canadian River compact negotiation permission reached the Senate on March 2 and was approved, allowing the Senate's study of the Canadian River Project to begin. In April, New Mexico senator Dennis Chavez helped guide the bill out of committee while Texas supported

New Mexico's desire for a small Bureau of Reclamation project on the Dry Cimarron tributary the Vermijo River. Now, opposition came from Kansas, Utah, and Arizona. Senator Andrew Frank Schoeppel of Kansas insisted that the bill be debated in the full Senate because "it involves a policy of creating a domestic water supply for municipalities under reclamation law." Utah's Arthur V. Watkins accused the Texans of wanting the federal government to provide municipal water, a precedent that A. A. Meredith denied. Meredith assured Utah's solons that Texans only desired "the credit of the federal government to make it possible for us to spread repayment over a period of time sufficient for us to meet this obligation." Senator Ernest McFarland of Arizona had a somewhat related complaint. He told Texas's delegation:

> I'm in real sympathy with this project with this one exception—that we have never had a complete repayment and this will set a precedent that our people in the West, namely in Arizona or Utah, can't afford to have set by reason of the fact that our people cannot repay the cost of their projects . . . and so we will oppose this project unless you reduce your repayment by half and let the Federal government pay half.

As George Finger later recalled, it took some persuasion, but the Texans got their legislation passed, even if they still insisted on paying the whole bill.[44]

On June 25, 1950, troops from the People's Republic of Korea (North Korea) invaded the South in order to forcibly unify the country by eliminating the elected leadership of the South. Five days later, President Harry S Truman ordered American troops to enforce the United Nations' calls for North Korea to remove its troops from the South. American military reservists were called to duty, troops and supplies began heading for Japan and Korea as fast as possible, and the Canadian River Project suddenly became one of the least critical items on the Senate Appropriations Committee's agenda. Funds for any sort of dam and aqueduct disappeared. In order to keep the bill alive, Johnson and Connally agreed to another compromise: they promised to request no appropriation until after the "war emergency" ended. With that agreement, the Senate passed the bill on December 15, and on December 29, 1950, President Truman signed Public Law 898-81 into law. A. A. Meredith sent George Mahon a box of frozen pheasants in celebration.[45]

The cities of the Panhandle and South Plains now had all that they needed to build their water project except for funds and a river compact.

Wartime and lingering questions about "wartime emergency" muted the celebrations attending the signing of the authorization. How long was the war in Asia going to last? What conditions would New Mexico impose in exchange for agreement to a compact? As the next decade progressed, it would become apparent that obtaining the legislation for the Canadian River Project was far, far easier than forging agreements between eleven cities over payment for a dam and a pipeline. Even as representatives of Lubbock, Borger, Amarillo, and Plainview dickered over details, rolling dust storms returned, driven by drought worse than that of the 1930s. Once again, the environment held the trump card for all those living on the Llano.

Chapter 5

THE PAPER RIVER AND A LOT OF DAM TALK

Amarillo's mayor Gene Klein and the Canadian River Project Organizing Committee's secretary, A. A. Meredith, may have occasionally wished that the High Plains' most famous historian would go far away, especially on January 11 and 12, 1953. J. Evetts Haley protested loudly, publicly, and rather persuasively against their precious Canadian River Project, confusing area residents and hampering CRPOC's efforts to gain approval for the project from the Texas Board of Water Engineers, the body that apportioned surface water in the state. Haley remained firm: he opposed the dam and would do all in his power to stop it unless Klein, Meredith, and the others "fired" the Bureau of Reclamation and built the dam with private money and labor or persuaded the Army Corps to build it instead. Meredith, Klein, Lawrence Hagy, Clarence Whiteside, and other dam proponents began to realize that as hard as it had been to obtain the legislation authorizing the Bureau to build the dam and aqueduct, those challenges might be minor speed bumps compared to getting the project under way. Even the pressure of a drought worse than the 1930s did not speed the process.[1]

Before any VIP turned the first shovel of dirt for a dam, people across the region haggled over and carved up the "paper river." As they argued, they proved once again that there are few things as combustible as the combination of water and money. First, Texas, Oklahoma, and New Mexico—and the federal government—divided the river's waters. Then the eleven cities interested in using that water formed a legal body with the power to sign a contract with the Bureau of Reclamation. Once the states and the U.S. Senate approved the compact and the municipal group organized and achieved state recognition, disagreements over dividing the water and paying for the redirection of the Canadian's flow to Lubbock and Lamesa took the greater part of the decade to resolve, a fact that placed the Canadian River in the company of the Colorado River and other well-litigated streams. Regional and national tensions that had been submerged during the efforts to obtain authorization for a Canadian River project shot to the surface once the interested towns faced paying for the dam and aqueduct. Towns voted themselves in and out of the project, accusations of greed appeared in print, and finally eight of the eleven cities tried to finance and build the project on their own, failing only because of a lack of water buyers. A crippling drought, in some ways worse than that of the 1930s, added to the urgency in the High Plains and indeed in the entire state of Texas and much of the nation. The 1950s were anything but "Happy Days" for those trying to build a reservoir in the Canadian River Valley.

• • •

Before a single clump of dirt got moved or pen glided over paper, prospective water users in Texas needed to negotiate a river compact with Oklahoma and New Mexico. A sense of urgency pushed the men from the three states and the federal government: without the compact there could be no contract for construction of the water project, and Texans wanted the dam built immediately. The initial meeting in 1948 produced no results because the governors had not yet appointed official negotiators or commissioners. The representatives from Oklahoma and New Mexico expressed interest and encouragement, but could not make any potentially binding agreements or statements. Informal contacts continued and in 1951, following passage of the enabling legislation, negotiation began in earnest.[2]

The question arises: why would New Mexico, upstream of both Texas and Oklahoma, be so concerned about what happened to water that had left the state? The concern arose because under the current legal systems, Texas and Oklahoma had rights to the river's waters and could go to court in order to claim them. New Mexico needed to protect its share of the Canadian River, which meant codifying exactly how much water it could keep and how much needed to pass through the state and when. The legal battle over the waters of the Pecos River and earlier disputes involving the Rio Grande put New Mexico on the defensive, even though no one downstream of New Mexico used the Canadian for much besides watering livestock or irrigating a very small patch of fodder crops. The matter boiled down to fairness and legal equity.

One of the first water conflicts to come before the U.S. Supreme Court, *Kansas vs. Colorado*, established the doctrine of equity in water-compact administration. Water equity meant that if the upstream state benefited from water use but the downstream state did not do as well due to upstream diversions, and if the upstream economic benefits could not be spread over the entire portion of the river in question (by prosperous upstream farmers buying groceries and equipment from people downstream, for example), then the downstream state had firm legal grounds for forcing the upstream state to release water and to adjust water use accordingly. It would be relatively easy for Texans in the Canadian watershed to argue that New Mexico's retention of "excess" water deprived them of the waters needed to fill a dam and supply the cities. As a result, the more clearly the compact defined and sorted everyone's exact rights, the better protection New Mexico would have if and when it decided to build more reservoirs or to enlarge Conchas Dam. So New Mexico had great interest in a downstream dam because of the need to divide the waters in order to fill that dam. Thus New Mexico's insistence on having a compact before any construction could begin in Texas or Oklahoma.[3]

The compact needed to accomplish several tasks. First, it defined such things as the Canadian River and North Canadian River, "conservation storage," and other water-related legal elements. Next it allocated how many acre-feet of water each state could keep and how many acre feet had to go through to the next state downstream. The compact also included a way of negotiating later situations and difficulties. The provisions of the original 1928 Canadian River Compact concerning mutual assistance in construction of water projects were dropped and other things more clearly defined because of changes in federal funding and regional politics. As New Mexico water historian Ira G. Clark points out, unlike the earlier Pecos River and Rio Grande compacts, the Canadian River Compact emphasized conserving water for future regional growth, not on dividing up an already overcommitted stream. The resulting compact was succinct, perhaps too much so.[4]

The Canadian River Compact as negotiated in 1950 and signed into law in 1952 was short and to the point. Any water rights already in place remained unchanged. New Mexico could have all the waters "originating above Conchas Dam." It also had full use of "all waters originating below Conchas" and was allowed to store up to 200,000 acre-feet (a/f) of those waters. Of the North Canadian River, the compact limited Texas to more or less riparian uses: domestic use, livestock watering, and irrigation of lands "which are cultivated solely for the purpose of providing food and feed for the householders and domestic livestock actually living or kept on the property." Texas could keep 500,000 a/f of the main Canadian

River until Oklahoma constructed dams sufficient to store 300,000 a/f, at which point Texas could have the same amount as Oklahoma's storage plus 200,000 a/f more. Oklahoma kept everything that crossed its border and had the option to ask for more if Texas did not release all of the flood flows. As retired U.S. Representative George Finger explained later, "Oklahoma has the right to call for some release of water from the project, but has assured us that because of the sands in the river [bed] that they never will because the water would never reach the Oklahoma line." The Canadian River now began downstream of the Conchas Dam spillways as far as Texas was concerned. Oklahoma's river began at the proposed Texas dam, relieving the state of many of its former flood sources. A four-member board comprised of representatives from New Mexico, Texas, Oklahoma, and the U.S. government would meet periodically to discuss matters, to approve any changes, and monitor the construction of smaller dams within the watershed. All decisions would have to be unanimous. The working group finalized the Canadian River Compact on December 6, 1950, and all seemed in order and smooth. The legislatures of the three states passed the compact by May 10, 1951. The U.S. Senate ratified the Canadian River Compact and it became a legally binding contract and treaty on May 17, 1952.[5]

Meanwhile in Texas, the cities interested in obtaining the newly available water from the Canadian had to form a new legal body. The old Panhandle Water Conservation Association no longer fit the needs of the municipalities, nor did it have the legal or financial power necessary to negotiate with state agencies or with the federal government. As a result, the Canadian River Project Organizing Committee had come into being June 17, 1949, comprising appointed or volunteer representatives of the various cities and towns interested in obtaining Canadian River water. In many ways this was a placeholder organization, created with the understanding that a different body would take over its functions once official recognition via state legislation took effect. Once formed, the CRPOC took rapid steps to guarantee access to any Canadian River water eventually impounded by the dam. In 1952 the CRPOC filed a water claim with the Texas Board of Water Engineers (TBWE), requesting an allocation of 150,000 acre-feet per year on the Canadian River. Although only five claims to the Canadian's waters existed in Texas, three of them upstream of the dam site and all of them small claims for domestic or nature preserve use (Rita Blanca Lake, for example), the CRPOC did not care to find that someone else had placed a prior claim at the last minute that they would have to buy out. As their request joined the long list of petitions for the TBWE to consider, High Plains and South Plains residents turned their attention to a more familiar problem: that of obtaining the cash needed to build any dam and aqueduct.[6]

Although they had the compact in hand, funding the reservoir and aqueduct remained another major hurdle for the CRPOC to clear before work could begin. The project authorization bill had passed the Senate on the condition that the Texans made no funding requests until the end of the war emergency. As 1951 progressed, Meredith, Hagy, Johnson of Lubbock, and others decided that the conflict in Korea no longer constituted an emergency and that the time had come to begin asking for appropriations. However, Senators Lyndon Baines Johnson and Tom Connally did not share their opinion. A delegation traveled to Washington in late August 1951 to talk with George Mahon (House Appropriations Committee), Lyndon Johnson, and Tom Connally about starting the appropriation process, but the senators rebuffed them. According to the report in the Fort Worth newspaper, the senators were as blunt as the Panhandle delegation was unhappy. According to the solons, Amarillo and the other members of CRPOC would just have to wait their turn for consideration of the funding request.[7]

Back in Texas people discussed the efficacy of arguing that the dam constituted a security and national defense item because it would provide water for the air force bases in Amarillo and Lubbock as well as relieving pressure on irrigation farmers. After all, adequate food and fiber constituted a national security concern. Since none of the eleven cities currently suffered from an immediate lack of water and neither did the two bases, the national security argument failed to have any effect and was dismissed almost as quickly as it was created. According to the powers in Washington, the dam builders just needed to be patient and to work on those things that they could sort out themselves. And there were plenty of matters that needed to be dealt with before the Bureau could take further action.[8]

According to state and federal law, the Bureau and the cities had to sign a contract for repayment before any work on any project could begin. However, the eleven cities could not individually contract with the Bureau, and the State of Texas recognized neither the CRPOC nor the Canadian River Municipal Water Users Association (CRMWUA), the first successor to the CRPOC, as taxing and regulating entities. The contractor had to be able to collect taxes or fees in order to have revenue with which to repay the Bureau of Reclamation. The old Panhandle Water Conservation Association could collect dues but lacked taxing power and had not focused either on the Canadian River or on municipalities. This lack of taxing power and state recognition did not stop the CRPOC from placing a water-rights request with the state Board of Engineers for the flow of the Canadian upstream of Sanford, Texas, however. They filed the request in order to initiate engineering studies and with the understanding that as soon as a legal body existed, it would replace CRPOC on the state paperwork as the water-right holder. But water rights and understandings still did not allow the CRPOC to sign any sort of contract with the Bureau.[9]

In 1953 State Representatives Morris Cobb of the 123rd District and Waggoner Carr from Lubbock introduced authorizing legislation into the Texas House. A new legal entity, the Canadian River Municipal Water Authority, would replace both the CRPOC and CRMWUA. CRMWA (pronounced "krim-wah") could sign contracts with the Bureau of Reclamation for construction of the Canadian River Project. It would hold the rights to all the Canadian River's water, after the four small senior projects upstream got their share. CRMWA would also have the power to issue revenue bonds and to charge member cities for the water that it provided from the river. CRMWA would then use those monies to repay the Bureau of Reclamation, to maintain the dam and aqueduct, and to pay a small staff. The various member towns and cities would keep their own utilities departments but would contract with CRMWA for all or some of their water needs, and would have one member on the board for each one hundred thousand people in the town or city. And neither Amarillo nor Lubbock could hold the position of president of the board, in order to limit their power.[10]

As Morris Cobb explained later, the bill faced opposition. The entire Canadian River Project and its various parts would set precedents that displeased some downstate legislators. "We had opposition from legislators in north and east Texas [who wanted] to zealously guard all of the water that flows into or through their territories and their streams and nobody, even today, wants to give up water rights because water is such a basic commodity," Cobb told historian Debra Wood in a 1990 interview. Although the Canadian did not flow through any more of the state than the Panhandle, this would set a precedent for other river basin associations, such as the later Colorado River Municipal Water District (CRMWD). In addition, the state had just gotten through the Tidelands dispute with the federal government over oil rights, and downstate legislators worried about the Bureau's claims to the Canadian becoming the proverbial camel's nose into Texas's inland waters. Panhandle legislators pointed out that the water rights would belong to CRMWA and the state, not the Bureau, a fact that soothed some concerns. Opposition also came from outside the state because of the enormous cost of the pipeline. Eighty-five million dollars was a lot of money in 1953, especially to spend on eleven little towns in a state without any federal land. This did not sound like "reclamation" to some opponents of the project.[11]

Despite opposition, both the legislature and the state Board of Water Engineers eventually approved of the new organization. On March 11, 1953, the board granted CRMWA the right to organize and allotted it 100,000 acre-feet for municipal uses and 51,200 acre-feet for industrial uses, per year, and gave permission for CRMWA to retain up to 961,000 a/f total storage from any remaining unappropriated waters of the Canadian. Since only Rita Blanca Lake, two small pasture irrigation works, Lake Marvin downstream of the dam, and a domestic claim held senior appropriations on the river, the engineers' ruling meant that CRMWA could take the Canadian's entire flow, up to 150,000 a/f per year (assuming that much water did flow in a year).[12]

The final legislation authorizing the Canadian River Municipal Water Authority passed in 1953. The "Conservation and Reclamation District" included (subject to approval by the voters): Pampa, Borger, Amarillo, Plainview, Littlefield, Lubbock, Tahoka, Lamesa, Brownfield, O'Donnell, and Levelland and those businesses and households that depended on the municipalities for water. An appointed general manager and an elected board of directors ran the Authority. Each city elected one director to the board, and those cities with a population larger than ten thousand people chose two directors. The cities and towns in the District could contribute money for engineering studies and other projects necessary for the construction of the dam and aqueduct. The District, as described in section 13 (a), was authorized:

> To store, control, conserve, protect, distribute and utilize within or without the District or within or without the state the storm and floodwater and unappropriated flow of the Canadian River and its tributaries, and to prevent the escape of any such waters without first obtaining therefrom a maximum of public benefit, by the construction of a dam or dams across said river and its tributaries, or otherwise, by complying with Chapter 11, Water Code, and in such manner as shall fully recognize and be in harmony with the limitations of use of the waters of said river provided in the "Canadian River Compact" appearing as Chapter 43, Water Code. The District is also empowered to provide by purchase, contract, lease,

gift, or in any other lawful manner, and to develop all facilities within or without the District or within or without the state deemed necessary or useful for the purpose of storing, controlling, conserving, protecting, distributing, processing and utilizing such surface water and the transportation thereof to the cities and areas comprising the District for municipal, domestic, industrial and other useful purposes permitted by law. [As amended Acts 1987, 70th Leg. ch. 251, par. 2, emerg. eff. May 28, 1987.][13]

In addition to tapping the Canadian, the District could also, if necessary, according to text added later, "acquire and develop within or without the District or within or without the state any other available source of surface, storm, flood, underground, or other water supply and to construct, acquire and develop all facilities deemed necessary or useful with respect thereto." [14]

In order to pay for the main cost of the aqueduct, dam, and other parts of the water system, the District could borrow money and issue bonds. The towns remained free to float their own bonds for civic projects, including separate water and sewer projects, but they would also be responsible for their share in the District's debt, forming a parallel series of water-related bond issues. The legislators assumed in 1953 that all the money to be borrowed would come from the federal government. As with all such state-approved water districts, the designation made provisions for amendments and revisions as needed. The legislation became law in May 1953, and the only thing remaining was for the residents of the interested towns and cities to voice their desires through a special election.[15]

Proponents of the water project labored hard to convince voters that the area's current and future survival and prosperity depended on approving CRMWA and the project. Shortly before the November 24 election in Lubbock, for example, long-time project supporter Clarence K. Whiteside published a detailed question-and-answer pamphlet about the proposed organization and project. Why should voters agree when there was not yet a definite plan, engineering, financial, or otherwise? Because, he wrote, officials in twelve towns have been at work, via the CRMWUA, for five years and "there are plans," both with CRMWUA and the Bureau of Reclamation. He assured readers that the cities closest to the dam would not pay lower rates, that it would be feasible despite the costs and long aqueduct required, and that nothing was wrong with Lubbock's well fields. Whiteside also admitted that the project would *not* meet the city's needs for the future, but only for ten years, then it would produce declining returns for the next forty or fifty years. He argued that "[w]ithout the Canadian and/or proration we would not need nearly so much [water] for the reason enough golden gooses (irrigated acres) would already have been killed to the extent that the city's demands would be much smaller." On the next page he repeated Lubbock's mantra, under-lining it for greater emphasis: "A gallon of water from the Canadian is a gallon saved for irrigation." Even at a price between twelve and fifteen cents per thousand gallons the water came at a lower economic and social cost than new wells would.[16]

Irrigation saved the economy of the South Plains during the drought of the 1950s, but it also put the city of Lubbock into an uncomfortable bind. Since the 1930s, cotton had become the king of the region's economy. A new type of long-fiber, upland cotton and the development of mechanical cotton-pickers, when combined with plentiful and inexpensive ground-water, helped boost the area's fortunes. The boll weevil that destroyed much of the Deep South's cotton had thus far missed the Llano Estacado, making the land even more attractive for cotton growers. The fluffy white fiber became one of the three foundation stones of Lubbock's economy, along with Texas Tech University and government functions (state and federal). But Lubbock and the cotton growers collided over water.[17]

The Sand Hills well field in Lamb and Bailey Counties west of Lub-bock formed the flash point. In the early 1950s the city purchased water rights under the sand hills where irrigation seemed impossible and the only competition for water came from ranch windmills. The water "lay fallow," so to speak, until 1955, when Lubbock faced rapidly declining wells within the city limits and close to town. It was time to begin bring-ing water into Lubbock, using a pipeline that would run along the high-way from Muleshoe. Cotton farmers and others around the proposed well field responded negatively and loudly. Lubbock, one letter phrased it, would be "the scarlet whore of Babylon," stripping the land of its water and prosperity for her own suicidal benefit. Challenges arose to the wells, to the pipeline right-of-way, and to almost every other aspect of the project. Changing irrigation technology now made it possible to grow cotton on the edges of the sand hill area. There were protests that if Lubbock went afield for water, what would stop Big Spring, Abilene, or other towns from also stripping the area of its water. These last objec-tions skipped over Amarillo's 150-mile (241 km) water reach, but the point remained. Amid calls for boycotts, heated letters, and fierce opposition, the city's leaders eased off on their timetable.[18]

Part of the city's problem stemmed from the drought. There were three differences between the "Dirty Thirties" drought and that of the

"Filthy Fifties." First, the latter was worse—less rain fell in the period 1953–58 in the plains of Texas than fell in the 1930s. Second, the dust storms were smaller because more land remained under grasses native and domestic. Third, ranchers and farmers now had oil royalties to help them. In response to the glut of the late 1920s and early 1930s, the State of Texas passed legislation instituting proration, or production rationing. Oil producers no longer had to pump all their oil at once, because pumping rationing allowed oil to stay in the ground without threat of someone else's well draining it. New discoveries in the South Plains benefited a number of ranchers in that area as well. Instead of depending on the federal government for hay relief, ranchers could use their oil royalties to pay for imported cattle feed or for moving their livestock to greener pastures, thus protecting their own lands. Irrigation allowed farmers to make some sort of crop even if heat stress reduced yields. The drought covered much of the country at various times, leading cities all over the United States to declare drought emergencies at one point or another. But the national economy boomed, preventing a repetition of the human misery of the 1930s. The thirsty grass and trees suffered as much or more than twenty years before, but the drought did not impress itself on the nation's awareness the way the 1930s drought did.[19]

Meanwhile, as CRMWA proponents lobbied voters for the project, opponents mustered their arguments. One unexpected voice of caution came from Howard E. Robbins, the former division head of the Bureau of Reclamation in Amarillo, now with the Veterans Administration in Lubbock. In a letter to T. E. Johnson and the *Amarillo Globe* and the *Amarillo News*, copies of which he also sent to Lubbock and other towns, Robbins stated that he wanted not to cause discord but to provide information. In short, no runoff fed the Canadian River anymore. The dam at Conchas had thus far failed to provide sufficient water to the Arch Hurley Irrigation District in New Mexico to fulfill the dam's obligations to irrigators. Not enough rain and snow fell to meet the predicted water supply. Information from the Bureau "has been optimistic and promotional, interwoven with politics in an attempt to justify maintenance of their personnel in Amarillo and with out question, there is no certainty of a water supply that could guarantee the people of Borger, Pampa and Amarillo of a water supply of any consequence, much less all the other towns." Robbins continued with a warning about the necessary taxation and the problems that Dallas, Fort Worth, and San Angelo were already having. He concluded that each town should have its own separate groundwater supply system and that city leaders needed to work with the Department of Agriculture to begin conservation of the water and soil of the area. The project's other vocal opponent differed with Robbins over groundwater versus surface water, but opposed the project just as strongly.[20]

J. Evetts Haley remained the other major dissenter from the Canadian River Project. He protested on several grounds. As he explained in a detailed summary of his position, he did not disapprove of tapping the river or of the prospect of municipal elections for the project. His greatest concerns centered on how the campaign had been run by dam proponents and on the extension of federal power into the region, especially taxing power. Haley argued that most of the benefits would go to Amarillo, giving it "a pleasure resort," and to Borger ("a potential source of future industrial development") while the rest of the area would end up paying more than $232 million in basic cost plus interest over the next fifty years. He reminded readers that R. L. Oldham, Lubbock's head of public works, had assured city residents only one year before that the Sand Hills field in Lamb and Bailey Counties would provide 650,000 acre-feet, enough to meet Lubbock's needs for the next fifty years. Haley also pointed out that the argument for "saving" the Sand Hills well field made no hydrologic sense because the water already leaked out at the surface as well as flowing southeastward underground.[21]

Haley objected strenuously to the federal government's role in the proposed project. Recall that Haley had seen the costs of federal projects as well as the benefits during the drought of 1932–36, and he knew how federal projects seemed to creep and expand past their initial stated limits. Haley pointed out that the language of Public Law 898 gave control of the project to the secretary of the interior. "Its one sided nature providing for Federal control is clearcut [sic] in its further provision that the dam shall not be constructed until 'the repayment . . . is negotiated . . . by contract'—not satisfactory to all of us—but 'satisfactory to the Secretary.'" The law placed no limit on costs and required the cities to pay whatever the secretary declared, while the project remained under the secretary's control. This, Haley warned, was a very poor arrangement indeed and one to be avoided.[22]

Haley was not inflexible, unlike Robbins. The historian-rancher reportedly told Gene Klein and others that if they wanted to finance the dam with their own money, or to get the Army Corps of Engineers to build it, then he would cease his opposition. That would eliminate the "problem" of creeping control by the secretary of the interior that so worried him. "Remember the Tidelands," Haley warned, echoing the legislators whom Representative Cobb had quoted in his account of obtaining the CRMWA legislation. The federal government had already tried to take control of Texas's natural resources once before. Then it had been oil, Haley cautioned. Next it would be Texas water that the secretary locked up. A few individuals, including Lamesa City Council member Tracey Campbell, picked up Haley's complaints and wrote to George Mahon and others, but come November the majority of townsfolk decided that they preferred river water with a debt to the federal government over groundwater without debt.[23]

The ranchers who stood to lose property to the water project raised few if any objections, at least where they can be found by later researchers. The Masterson and Bivins and Whittenberg families, along with the Sanfords, owned businesses, ran cattle, and served on boards and in city government. And they knew that they would be compensated for the loss of land and water rights. No public land existed in the Canadian River Valley until well downstream of the Sanford site, and the valley's ranchers may have felt that the loss of ranchland cost less than they gained in goodwill for their sacrifice, flood control, and the opportunity to buy more land elsewhere. And while Mrs. Sanford had commented that a dam would ruin good rangeland, her heirs may not have seen things in quite the same way.[24]

As the year moved on and 1953 became 1954, J. Evetts Haley's suggestion that the cities turn to private funds for the dam became a more reasonable option, at least for some. Federal funds for the Canadian River Project remained unavailable despite repeated requests, in part because no contract yet existed between the organization now called CRMWA and the Bureau of Reclamation. The Bureau could not do any work without having a contract. As negotiations began among the different towns, some in the eleven cities argued that the municipalities should at least start considering the possibility of floating bonds for the project themselves—especially since discussions over exactly *how* the cities would divide the costs and pay for the project grew increasingly heated as more details emerged. Should billing be strictly based on a per-gallon division, so that Amarillo and Lubbock bore the brunt of the cost? Or would all the towns share in all the expenses? The residents and leaders of Amarillo, Borger, and Pampa did not want to pay for the aqueduct serving Lamesa, although they might be willing to pay a larger proportion of the dam based on their greater water use. Some suggested a two-tier payment plan, with part of the costs, including maintenance and administration and reserve funds, paid for on a per-gallon basis. The cities would then contribute to the infrastructure based on what share of the pipelines and pumps each town needed. Amarillo and the northern towns preferred to be billed along those lines, but the small communities at the south end of the pipeline were not as enthusiastic about the idea. However the commissioners worked out the payment system, some form of agreement had to be reached because until the Bureau had guaranteed buyers for a minimum amount of water, the Bureau would do nothing.[25]

However, other lenders existed besides the federal government. The Bureau of Reclamation had offered a fifty-year repayment plan and charged 1 percent interest, with a $3 million discount for the flood control and wildlife protection aspects of the project. Using these figures as a baseline, Lubbock and the other CRMWA members approached several private bond agencies to see what they would charge for lending the necessary money. The resulting bonds would be revenue bonds, meaning that the cities would turn over the income from water sales to the bondholder to pay back the debt. First the bonding agencies required CRMWA to pay between $65,000 and $70,000 for private engineering surveys because the companies did not trust the federal numbers. Once the engineers finished the surveys, the cities heard the bad news. Money for municipal projects was becoming expensive, and all the companies consulted wanted at least 3 percent interest with a thirty-year repayment period. And they demanded a higher minimum guaranteed water sale before agreeing to provide the funds for the project.[26]

The payment question reached a head in 1955–56. A landowner offered Amarillo's city government the opportunity to purchase more than one hundred thousand acres of groundwater rights in Dallam and Hartley Counties, roughly 120 miles (193 km) northwest of Amarillo. No one irrigated there yet because the rolling, sandy terrain made flood-furrow irrigation pointless. On March 14, 1955, city residents voted a breathtakingly large $6 million bond to buy the water for future use, 2,532 for and 580 against. The area lay in an untapped portion of the Ogallala Aquifer, uphill from Amarillo, so the cost of moving the water would be somewhat less than water from other sources. This relieved the city of some of the earlier pressure and helped Amarillo's CRMWA representatives stand the city's ground against, as the editor of the Lubbock newspaper phrased it, "a little prodding from some of their less [financially] cautious neighbors." At the same time, Pampa's city leaders began expressing some concerns about the cost of the water, which they did not really need yet.[27]

The differences in cost and payment between the private and federal options almost sank CRMWA. Before Amarillo's bond issue, CRMWA's board of directors voted to ask the cities to allow the Authority to float private bonds to finance a privately constructed dam. It would be smaller than the one proposed by the Bureau, and the interest rate would be greater by a percentage point with a shorter repayment schedule, but construction could begin as soon as the financing corporations issued bonds. As A. A. Meredith told Representative George Mahon, the area was at "the crossroads in our deliberations on financing" because Amarillo, Levelland, Pampa, and Plainview shied away from the proposed water cost, a price up to twice what they currently charged their customers. In response Amarillo asked the board to reconsider, which it did in May 1955 but decided to continue with the private bond issue. Amarillo's city commission debated the matter for eight hours and then, on July 28, 1955, at the first televised city commission meeting in Amarillo's history, the commission declared no. Amarillo had plenty of water and plenty of debt already. Amarillo's city commission voted to withdraw from CRMWA. Commissioner R. C. Jordan explained that the city had "an adequate supply of water for the next seventy-five years . . . and for a city of 200,000 people." The commission was still interested in a surface reservoir on the Canadian River, but not out of its own pocket. It is easy to imagine the chorus of groans and imprecations that greeted this statement in Lubbock, Lamesa, Tahoka, and O'Donnell. The loss was critical to CRMWA: Amarillo would have been the largest source of water revenues for the district. Now what?[28]

The southern cities and Borger remained undaunted even as Pampa and Plainview slipped away as well. Lamesa, O'Donnell, and Tahoka found no alternative—their wells were going dry and new wells produced only sand. Lubbock's representative to CRMWA, Robert Jordan, declared that this was a chance to regroup and perhaps shift direction; Amarillo's water share could be sold to industry. At subsequent meetings, Pampa followed Amarillo's lead for a similar reason. Pampa possessed sufficient groundwater for the moment, and residents did not care for the "pay by the mile" option. They were at the far end of the east–west pipeline through Borger and would have to cover roughly 80 percent of the cost themselves. Although the oil boom raised incomes and tax revenue in Pampa, they did not rise enough to justify the cost of the pipeline and higher water rates.[29]

River project supporters called two more elections: one for November 1955 and one for March 1956. Predictably, the November election on whether to issue bonds passed with an 11–1 ratio in the remaining nine member cities. All that remained was for the nine cities to approve the necessary water contracts, binding themselves to purchase a minimum amount of water in order to support $78 million in bonds. Then bonds could be issued based on those contracts and work started on the dam. But on March 13, 1956, Plainview threw a wild card into the game. The new federal highway missed Plainview, bypassing the town by a mile or so. Residents blamed the mayor, who had also been the main proponent for Plainview's participation in CRMWA. The town possessed excellent well water, and although there was increasing irrigation in the area, the small city's needs did not conflict with pumping by Hale County farmers. City residents launched a recall and ousted both the mayor and CRMWA, voting 5 to 2 to turn down the contract. A second vote in April produced the same result—Plainview would remain out of CRMWA for the foreseeable future. Plainview residents remained interested in a water pipeline but for the moment preferred to keep their money at home and drink well water, just like Amarillo's city commission members did.[30]

Lubbock, Borger, and the southern cities took the bit in their teeth, as it were. They needed the water, especially the towns south of Lubbock where the Ogallala Aquifer was thin. Eight towns signed firm contracts with CRMWA, agreeing to pay 13¢ per k/gal of Canadian River water. At the recommendation of the bonding companies, CRMWA "employed a New York firm who made the survey . . . at an expense of about $60 70,000 dollars only to find out that the bonding companies' rate of interest was about four percent on the estimate and the private engineers estimate was about $68 million but lacked any flood control features and also impounded almost a third less water." This was much worse than what bankers, including William Smallwood of Dallas, had suggested the year before. At the same time the remaining cities cast about for someone to buy enough water to make up the difference, preferably someone not in agriculture. CRMWA's members approached Philips Petroleum in Borger to see if the company might buy at least one-third of Amarillo's "share," the minimum necessary to start the process for private bonding. After much discussion and negotiation, Phillips named a final no-changes price that was 0.5¢ per k/gal too low to meet the sale minimum for the private financing agencies. The groans that arose in the Lubbock, Lamesa, and Tahoka city council chambers must have rattled the ceiling tiles.[31]

As Lubbock tried to placate its agricultural neighbors and woo Amarillo, Lamesa and Tahoka turned their eyes south. The Colorado River of Texas began near the southern edge of the Caprock and then flowed southeast toward San Angelo before turning toward the Gulf of Mexico. While residents of the Llano looked to the Canadian, people living in the hills south of the plateau had started drawing up plans to capture the waters of the Colorado. This might provide a solution for Lamesa's short-ages, although it would be expensive to pump the water uphill. However, the process that had been in place for six years for the Canadian was just starting for the Colorado, meaning they faced an even longer time lag before any water reached the thirsty towns.[32]

Forced to reach a compromise, the cities hammered out an agreement over the next eighteen months. In February 1957, the state legislature amended the enabling legislation to allow CRMWA to set separate water rates for each member. After more negotiations and the guarantee that the Authority would not apply for private financing, Amarillo, Pampa, and Plainview rejoined the group. Amarillo and Borger and Lubbock agreed to pay a greater proportion of the cost of the dam and its maintenance because they foresaw gaining the greatest benefit from the structure. The pipeline and pumps would be charged to the cities by the mile, meaning that the southern towns bore the brunt of the aqueduct cost. However, they did not pay for all of the pipeline: CRMWA would fold some of the expense into the overall costs billed. Each town had a minimum water purchase, set lower than originally proposed, but each town also retained the right to sell some of that water to other member cities. For example, if Pampa did not need all of its share, it could pass the extra to Lubbock if the board approved the purchase. The cities paid for the water and water treatment by the gallon, with Lubbock treating the water for the cities to its west and south. Amarillo caused a hiccough in 1958 by refusing to sign anything until the final water prices were announced. Once the Authority ironed out that bump a month later, CRMWA secured funding from Congress to move ahead on project plans. In the end, no one was completely happy, suggesting that they had reached the best compromise that could be worked out at the time.[33]

The long wait was over. Even with negotiations still underway, CRMWA collected $12,500 from its members and commissioned the Bureau of Reclamation to do an updated study in September 1957. After all agreements had been reached among the cities, CRMWA accepted the Bureau's report and plans. On November 22, 1960, residents of the eleven member cities went to the polls one last time, voting to accept or reject the Bureau's plan. They accepted, and on November 28 representatives from nine cities of CRMWA signed a contract with the Bureau of Reclamation. Slaton and Lamesa followed shortly thereafter. In exchange for $96 million to be paid over fifty years, the Bureau promised to build an earthen dam and more than 315 miles (506 km) of pipeline to provide up to 105,000 a/f per year to the participating towns and cities. The 3 percent interest charge would not start accumulating until the first water deliveries started, and would be waived if the project failed to supply more than 70 percent of the contracted water (i.e., until the dam was finished, reservoir filling, and not in a drought). The era of struggle had ended. Now the Canadian River could begin doing its duty to the citizens of the Llano Estacado, freeing them from reliance on dwindling groundwater supplies.[34]

Between 1950 and 1960 the Canadian River in Texas was dammed and divided, at least on paper. The Canadian River's flow belonged to CRMWA and the member cities, or would once they finished paying the Bureau of Reclamation for the as-yet-unbuilt dam. However, none of these negotiations, compacts, contracts, and disputes affected the "wet" river, which continued nibbling pastures, swallowing cars, and providing water for livestock. That is, when it ran. The 1950s saw the start of the space race, the development of the interstate highway system, jet airliners, and television. But for those living in and along the Canadian Valley, the decade harkened back to the 1930s, as red, black, and brown dust sifted under windowsills once more and ranchers sought ways to keep their cattle alive when the rains failed.

Chapter 6

The Wet River in a Dry Decade

C ows read neither newspapers nor legal documents. Therefore it is highly unlikely that the cattle and other animals grazing in the Canadian Breaks and drinking from, wading through, bogging in, and occasionally swimming across the river noticed many changes to the river in the period between 1948 and 1963. Cattle and horses grazed, mated, ruminated, and swished their tails at the summer's flies, all unbothered by war, legislation, or changes in the larger faunal assemblage of the Breaks. Deer, reintroduced to the valley, sampled gardens and helped themselves to irrigated pasturage, while wild turkey drove the few invited or lease-paying hunters to distraction as they hid in the thickening brush. Drought returned along with dust storms and floods. Blizzards chilled humans and livestock alike, sending cattle drifting south just as their predecessors had done in the late 1800s. Life along the wet river changed in many ways, but the daily and yearly round of flood and storm, heat wave and blizzard, in other ways remained unchanged.

The taming and subdividing of the paper river initially affected the wet river not one bit. Nor did it affect the river as people thought of it. Changes in the Canadian Breaks came in the form of electricity, more oil and gas wells and pipelines, newer cars and trucks, the further spread of mesquite and minor shifts in the drought-flood cycle. Outside of the valley, the 1950s saw the creation of the first environmental protection legislation aimed at preserving or improving the quality of air and water. But the Panhandle and South Plains, although economically dependent on the newly regulated extractive industries, seemed unaffected by most of these trends.

One of the changes that occurred in the Canadian Valley came to the far western end, in Oldham County. Stockholders in the Matador Land

and Cattle Company, headquartered in Dundee, Scotland, considered the value of their holdings and as a group voted to liquidate the company. The corporation broke up the component ranches, selling them and their associated mineral and water rights and ending more than sixty years in the cattle and land business. The entire Alamositas Division came onto the market as a single block, providing a rare opportunity to own a large, contiguous ranch with guaranteed water. A contractor from Lubbock, Texas, took advantage of the opportunity. J. R. Fulton had always wanted a ranch. The romance of the Old West and of ranching caught him at an early age, and now the oil-field pipeline expert had both the resources and the opportunity to indulge his desires. He did not hesitate, buying the land and all the records of every division of the Matador as well. Fulton donated the ranch records to Texas Tech University in Lubbock, forming one of the cores of the university's western history archive. The Alamositas Division became the Canadian River Ranch. Fulton kept the ranch a private, working ranch. Fulton's new ranch manager and other employees found themselves having to work with and around the Canadian's and Rita Blanca Creek's rises and vanishings just like all the other denizens of the Canadian Valley, including the young men at Boys Ranch "next door" to the Canadian River Ranch.[1]

The boys living at Boys Ranch on part of the old LX north of the Canadian also abided by the Canadian's whims. Without a bridge, crossing the river remained a bit of an adventure when the waters rose in spring or fall. Local rancher and businessman Julian Bivins had given former pro-wrestler Cal Farley 120 acres (49 ha) of the LX ranch in 1938 so that boys in trouble or without families could have a second chance. Just as James Stevens, former manager of the Alamositas, had done in the 1910s to 1940s, Cal Farley and other adults who needed to get to town either drove carefully across the riverbed, were towed across, or drove back roads to reach pavement in order to get to Amarillo and cross the river there. The isolation forced the boys to learn to rely on themselves, kept them away from possible trouble in town, and prevented trouble from reaching them. It soon became apparent that pulling bogged cattle, tending horses, and exploring the brushy riverbanks probably did as much as the parenting and schoolwork the ranch provided to help the young men get sorted out and set on the path to more stable lives.[2]

Farther downstream, life continued on the LX, Masterson, and other ranches. Rural electrification had brought a new source of energy to the valley, making many aspects of life easier for the human residents of the Breaks. But cows remained cows, and the unpredictable comings and goings of the Canadian kept things much as they had been fifty years before. Here, east of Amarillo, the valley began narrowing and the terrain remained too steep to work in cars or on motorcycles for the most part, forcing the ranchers to depend on horses still. But instead of driving herds of older cattle to Amarillo for rail shipment to cattle feeders in Iowa and Illinois, ever-larger cattle-hauling trucks carried young Herefords and Angus to the stockyard and thence to the new feedlots that sprang up to shorten the chain between ranch, cattle feeder, and market. Modern consumers preferred younger beef over the prime three- and four-year-old cattle that had provided the cuts and flavor popular in the 1880s. This preference changed business only a little—the Alamositas, LX, and other ranches had served as cow-calf operations in large part since 1914, with the weaned calves and yearlings belonging to the Matador traveling north to gain weight in Montana, Canada, or South Dakota before going to slaughter, while other ranches sold their cattle to Iowa and Nebraska cattle feeders. Now the cattle just moved to a closer feedlot in Kansas or on the plains north and south of the Canadian.[3]

If people living in the towns around the Canadian thought much about the river, their thoughts probably centered almost solely on how they could tap its waters. The entire valley as far to the east as Canadian, Texas, belonged to ranchers who did not always take kindly to uninvited visitors picnicking, hunting, or wading in the tributary creeks. The Canadian remained something that people crossed on a highway bridge or (until 1954) tried to ford at Boys Ranch if they attended the rodeo or had business there. Only six years after the battle for the dam at Tascosa ceased, the State of Texas not only built a highway bridge across two miles of Canadian River sand and banks but also paved the highway from Vega north to Channing. People looked forward to the day when, once all the cities got their acts together and the federal government built a dam, they could go fishing, swimming, boating, and hunting along the river's shores as the lake filled. Otherwise, most Panhandle residents probably did not think too much about the state and behavior of the river one way or another. There were more important developments to dwell on, such as the growing oil and gas business, and the return of drought and dust storms.[4]

The damper-than-average trend of the 1940s did not continue past 1951. Although there were dry years between 1941 and 1950, the decadal average of 22.5 inches (572 mm) per year remained almost an inch above the fifty-year average of 21.66 inches (550 mm) per year. Within the decade, good years alternated with and buffered against the drier stretches. Beginning in late 1951, drought stalked the next five years, culminating in the driest year then on record for the city of Amarillo: a mere 9.94 inches (252 mm) of rain in calendar year 1956.[5]

This pattern stemmed from the El Niño–La Niña currents in the eastern Pacific Ocean. A weak El Niño in late 1951 lasted only five months before turning neutral. So things remained until May 1954, when a strong La Niña pattern developed. For the High Plains, La Niña, a cold-water event off the coast of South America, meant that winter and spring tended to be drier than average with fewer summer storms, due to the jet stream and associated storms moving farther to the north. The reduced winter snow and cooler summer temperatures weakened the summer monsoon in New Mexico as well, causing fewer storms to come rolling off the mountains into the plains. Autumn tended to be a little damper on average than in an El Niño year, but the 1950s were not average. The entire plains region from southern Colorado almost to the Texas Gulf Coast suffered as drought lingered and expanded. Even normally humid eastern Texas suffered, and Panhandle residents gritted their teeth (and dodged flying grit) as dust storms returned to the region.[6]

The "Dirty Thirties" seemed to repeat during the "gritty" or "Filthy Fifties." Once again, strong dry cold fronts lumbered down over the plains, moving topsoil tens to hundreds of miles. And again, women fought back with wet rags, towels over windowsills, and lots and lots of sweeping and dusting. A better weather forecasting system allowed for some warning of the oncoming dust, but the experience remained almost unchanged for people in the way. Bonnie R. Cox, a former resident of Pampa, recalled sitting outside after school and feeling the air go still. The children had been dismissed to go home ahead of the dust, but her mother reached the school late. The little girl recalled a red wall bearing down from the northeast, full of soil from Oklahoma. Nellie Witt Spikes, a farmer and columnist on the southeastern edge of the Caprock east of Plainview, described sand storms that began at night on a hard north wind. When the storm finally died the next morning, only a few battered plants and stripped stems remained of her garden and flowerbeds. A young girl living on the north side of Amarillo named Opal Bascom recalled walking back from school in dusty darkness. Thelma Milburn of Dumas said that it was just like the 1930s because the sifting dust ruined any food left out uncovered. If there was anything good about the dusters of the 1950s, it was their comparative rarity compared to the previous drought's dust days.[7]

The default setting of the region's weather remained "dry." Dust from the High Plains had likely contributed to the dust storms reported in eastern Kansas in the 1850s and '60s and earlier. The sand hills north of the Canadian River on the Texas–New Mexico border and on the northern banks of the river within the western Breaks came from sand eroded by the river and whipped up by the prevailing southwesterly wind long before the arrival of Spaniards or Anglo Texans to record the events. Following a dry winter, February through April always meant dust. The sky turned brassy and hard looking, and soon distant houses or cars faded into a reddish-tan fog (if the wind was from the southwest). That was an average, normal sort of dust storm, the kind that irritated eyes and noses and caused difficulties for people but that also came fairly often and that faded away as afternoon became evening and the winds calmed. What made the '30s and '50s different were the black or red walls of churning dust that turned day to night. Were these dust storms part of the effects of global warming in the twentieth century?[8]

When and how often moisture reached the ground in the High Plains began changing in the late 1800s. As the planet warmed following the end of the Little Ice Age, rainfall patterns on the High Plains shifted their timing. Winters became drier as precipitation grew more seasonal, concentrating in late spring and early summer with a second slight increase in early fall. Although the amount of rain and snow remained connected to the El Niño–Southern Oscillation (ENSO), the modern seasonal pattern had become firmly established by the 1940s. All types of precipitation declined in the 1950s, snow and rain, winter or summer, then rose again as the Pacific's waters warmed with El Niño's return. Meanwhile, although it seemed to many area residents that the Dust Bowl was once again living up to its name, the dusters of the 1950s rarely caught national attention.[9]

Why did the storms of the 1950s fail to attract interest? More of the country went dry at the same time, and Central Texas actually suffered more than did the old Dust Bowl. The storms struck less frequently and remained more regional, so Washington, D.C., did not receive any Texas and Oklahoma real estate this time. At the same time, the nonagricultural economy of Texas, and of much of the rest of the country, boomed in the mid to late 1950s. The great oil strikes continued as the Permian Basin fields expanded, offshore drilling advanced into the Gulf, and more rigs, wells, and pipelines appeared in the Panhandle and South Plains. Without a Depression in progress to generate "Okies," photographers found no weary faces for dramatic pictures. Even ranchers who lost their grass and had to send cattle elsewhere or sell them now had oil or gas rents or royalties in many cases, or they could find work off the ranch in order to buy food for themselves and their livestock. Better soil management techniques also paid off, helping reduce the erosion. Crops withered and pastures turned tinder dry, but most of the soil stayed in place or at least remained closer to the farm.[10]

Not everything suffered equally during the dry years of the mid 1950s. Mesquite brush, an opportunistic native shrub, appears to have further expanded its range through the Canadian River valley and Breaks during these years. American Indians used the spiny plant for firewood and food. The hardwood burns hot when it is properly dried. Mesquite beans, which grow in pods of up to a foot long, are high in protein and can be sweet, notably those of the honey mesquite. When ground into flour, the beans provided a nutritious addition to drinks or flatbread or stews. Deer also ate the pods and the leaves, as did insects.[11]

Mesquite, like humanity, is a weedy species, meaning that it spreads easily and rapidly into available habitat. A phreatophyte (from Greek meaning "spring or water plant": a water-loving plant with roots that extend to the water table), mesquite sends roots very far down in search of water. Other roots extend outward from the trunk and produce a mild herbicide that inhibits the growth of competing species such as grasses. Fire provided the main check on mesquite expansion in the High Plains. Young seedlings are sensitive to fire, and if hot enough, range fires can also weaken or kill older bushes as well. Although fire did not race through the short grass steppes and the Breaks on a yearly basis as it seemed to do in the eastern tall grass prairies, fires burned often enough to encourage grasses and forbs at the expense of mesquite. Cooler, moister weather also favors native grasses and herbs, allowing them to grow faster and compete better against the thorny legume.[12]

Ranching plus drought inadvertently boosted and accelerated the spread of mesquite into the Canadian Breaks. Ranchers loathed seeing their next year's income and fences burn up and fought any fires that they found. The histories of the XIT, LS, and other ranches include accounts of the heroic efforts made to stop advancing flames. The ranchers made increasing progress over the course of the early twentieth century, reducing the numbers and extent of range fires considerably from what had been the regional "normal." At the same time, intense grazing in some parts of the Breaks reduced the amount of grass available during dry periods. Cattle then turned to mesquite pods. The pods are palatable, especially the sweeter varieties, and as long as cattle do not overfeed or eat the pods without additional feed for too long, bovines can survive on mesquite. In the process, any undigested seeds are deposited away from the parent brush in a moist pat of fertilizer. It would be hard to imagine a more auspicious start for a seedling's growth and development. Drought further favored the mesquite over native grasses because the shrubs are more tolerant of heat and lack of moisture than are most grasses, allowing the brush to take advantage of the conditions and expand its range.[13]

Ranchers did not intentionally encourage mesquite. Like many unfavorable results of human activities, the expansion of aggressive species came about due to a combination of things: drought, fire suppression, overgrazing of native grasses, cattle eating mesquite pods. By the 1950s mesquite "infestations" became a matter of serious inquiry, and research and state agricultural agencies in Texas, New Mexico, and other western states began offering suggestions for mesquite control and eradication. In far South Texas, Robert "Mister Bob" Kleberg of the King Ranch developed a special combination blade and digger that mounted on the front of a bulldozer and ripped the shrub up by the roots. Extension researchers tested and recommended several herbicides, as well as very carefully controlled burns and if the patch was small using chainsaws to cut down the offending plants before treating the stumps with herbicide or burning the area. All these "solutions" cost money and were of limited use on exceedingly rough and broken terrain. By the early 1960s, the greens and browns of the grassy Breaks had been greatly augmented, if not entirely replaced, by the green of mesquite.[14]

Along with mesquite and domestic cattle, another invasive species had begun appearing in the Canadian River watershed during the mid-twentieth century. An unrecorded botanist or gardener introduced a quick-growing, feathery tree with lovely pink blossoms to North America in the early 1800s, and railroads brought the plant to the banks of western streams and rivers to serve as a barrier to erosion. The tree grew rapidly, spreading along watercourses. Indeed, it held stream banks in place. "Salt cedar," the common name for *Tamarix sp.*, marched along the watercourses of the west, crowding out native cottonwoods, hackberries, willows, and other riparian vegetation. It also drank prodigious amounts of water and expanded until by 1914, for example, salt cedar filled more than 600 acres (243 ha) of McMillan Reservoir on the Pecos River in New Mexico. The aggressive phreatophyte formed nearly impenetrable stands, choking streams and in some cases actually lowering the local water table by sucking up so much water. Native animals and birds shunned the dense thickets of salt cedar because the clumps were too tightly packed for most creatures to shelter or nest in and the plants' seeds do not appeal to native seedeaters. Although it was a nuisance in the 1950s, by the end of the century High Plains residents viewed it with a loathing rivaling that reserved for tax assessors as salt cedar grew to form thickets along the Canadian so dense that it even crowded out native grasses.[15]

The Canadian River valley was not the only watershed to be graced by salt cedar, and the case of neighboring New Mexico is instructive. Salt cedar moved quickly into the Rio Grande and Pecos River basins, crowding out native plants and possibly contributing to declining groundwater tables. Because reduced flow of the Rio Grande and the Pecos River affected the international and interstate agreements for water sharing, salt cedar infestations became more than just a nuisance. People discovered that simply cutting down the offending flora was counterproductive—it sprouted from the stumps, and any stems that washed up on the banks downstream rooted, growing into new trees. The Bureau of Reclamation undertook experiments on Caballo Reservoir in 1958–59 in an attempt to find what would work. Aircraft sprayed herbicide on part of the salt cedar thicket. Once the foliage died, workers with brush-cutting bulldozers moved in, ripping out as much of the salt cedar as they could, then burning the remains. Afterward workers sprayed the area with a follow-up herbicide treatment to discourage resprouting. The treatment worked in the short term, but the hard effort and cost probably discouraged some observers. Others may have wondered about the wisdom of spraying herbicide on a municipal water source, even though a great deal of dilution took place between McMillan Reservoir and the various towns' water intakes. River managers at Conchas Lake on the Canadian and farther east along its banks no doubt read of the Bureau's efforts with interest, even though they did not have major problems with the plant—yet.[16]

If a person were to visit the Canadian Breaks during the pre-dam days in Texas, what would he or she have seen? Depending on exactly where the visitor stood in the length of the valley, he or she would find a sandy riverbed between 100 and 300 yards (90–275 m) across. The river itself would not be that wide of course, because part of the riverbed width stemmed from the great floods of 1893, 1904, and 1941. It would take several more decades for the channel to narrow again, assuming another flood did not tear out the reformed banks. If it was late summer or winter, a mere trickle of muddy water would snake across the sandy flats. A summer storm or monsoon-season shower still would bring rapidly rising water with quicksand in its wake.[17]

Let's say the imaginary river visitor to an Oldham County ranch turns from the Canadian proper and decides to follow the small trickle of sweet, clear water flowing into the main channel from between some brushy plants. He climbs a shin-high cutbank that provides access to the old floodplain. Where it was formerly covered in mixed grasses such as little bluestem, now invasive sandburs compete with shortgrass and some yucca and mesquite for space on the old river flat. An old trail, perhaps a cow path or simply a shortcut made by a ranch manager's children, offers a way to follow the little stream up into a side canyon after the visitor crosses half a mile of sandy, slightly rolling ground. At the head of the stream he finds a pool and spring where cold water seeps out between rock layers. Downstream of the pool, a flat, marshy area hints of a long-abandoned beaver or muskrat pond.

Suppose the ranch visitor knows a little about the area's waters. He might wonder at finding a live spring here, even after the half-decade of hard drought. After all, springs south of here, in Deaf Smith County and down toward Lubbock, have all gone dry. Some wells also have failed, forcing farmers and towns to drill deeper to reach the plentiful water. The ranch manager had explained over breakfast that the springs weakened and one smaller one had gone dry during the worst of the drought, but all the old waterholes remained pretty wet. The manager thought the farmers had killed the springs with all their wells, but he allowed that the drought also could have done it. Things were worse down south, or so he had heard. He had also cautioned his guest that the spring water was good but cows drank from it sometimes, along with deer and other animals, so he might not want to drink downstream of the headspring. The visitor takes the cowman's advice and looks around as he finishes filling his canteen. The deep footprints of cattle in the mud remind him that cattle have the right-of-way, especially now that the cows have young calves with them.

The hoof prints lead up toward the spring. The ranch visitor decides to continue along the small canyon and follows it past the spring, scrambling up a scree-coated slope. Yucca, mesquite, and few other hardy plants cling to the loose ground before yielding to grass at the top of the hill. From here he stops and lets the south wind cool him off as he looks down over the valley. Behind him, grass and brush extend over the rolling, treeless terrain of the Breaks. The canyon turns into a fold, twisting up toward the elevation of the Llano Estacado's flat surface. To the left the ranch visitor notes a dirt-walled cut and walks over, inspecting the arroyo. The ranch manager had told him about how some of the folds turned into pasture-devouring washes during the 1930s that deepened a little in the 1950s. Looking back toward the valley, the visitor spots the cluster of cottonwood trees marking the ranch headquarters and begins the long trek back down into the valley proper.

Farther to the east, in Potter and Hutchinson Counties, the inner and outer valleys become one as canyon walls tighten moving closer to the river. The valley bottom flats now abut the walls, rather than rising into sand hills. This is where a dam will go, and beginning soon, or so everyone hopes. Father east the valley opens again, no longer pinched and confined.

This is where the greatest changes in the Canadian came. After 1961 this part of the Canadian Valley would never be the same, at least not in most humans' timeframes. Here the paper river and wet river would converge to form a lake and a second watershed of sorts, as Canadian River water began making the long, one-way trip to the South Plains.

Chapter 7

BUILDING A DAM TIMES TWO

T he guests and officials at the Canadian River project groundbreaking on June 30, 1962, felt as if they were standing in a barbecue. Bright Texas sunlight reflected off the rocks around them and poured down on their heads. George Mahon later wrote to H. S. Hilburn, editor of the *Plainview Herald*, "You missed a wonderful opportunity to get a suntan" on that baking, sunny, nearly 90-degree Fahrenheit (in the shade) summer day. Wes Izzard, *Amarillo Globe-Times* newspaper reporter and radio broadcaster, served as the master of ceremonies for the historic moment. The dignitaries he introduced included Representatives George Mahon, one of the driving forces in Washington for the project since its inception in 1948, and Walter E. Rogers from Pampa. Senators John Tower and Ralph Yarborough joined Floyd Dominy, the colorful and controversial head of the Bureau of Reclamation, Assistant Bureau Director Grant Bloodgood, local Bureau manager Lon Hill, Secretary of the Interior Stewart Udall, and a representative from the Texas Water Commission. Before the first charge of dynamite sent red dirt and creamy rock cascading to the valley floor, Reverend Jason G. Glenn of Borger's First Presbyterian Church offered up an invocation. Amarillo's Tascosa High School band provided music appropriate for the monumental occasion as the Panhandle sun baked the well-dressed observers and participants, some of whom probably thought fondly of the day when they could take a cooling dip in the as yet unnamed lake.[1]

The period 1961 to 1970 seemed to be one of progress and harmony, at least where development of the Canadian River was concerned. Elsewhere around the country, the goals of traditional conservation ran headlong into the new ideas and conflicts that increasingly marked the latter part of the decade. While the 1950s had been a period of argument over apportionment of the Canadian River, the subsequent period witnessed apparent municipal harmony and the damming of the river in both Texas and New Mexico, events that began the next phase in the river's "life." Across the western parts of the United States, increasingly fierce battles over dams and rivers pitted members of the environmental preservation movement against the supporters of the old conservation ethic. In some ways the dams at Sanford and Conchas marked the last achievements of the Progressive Era conservationists as new ideas about society and the environment became the subject of public debate and protest, with the Bureau of Reclamation standing in the middle of the fray.

Even before Lamesa's residents voted themselves into the new water system in September 1961, the Bureau began obtaining additional funds from Congress and purchasing land for the dam, reservoir, and aqueduct. President Eisenhower had presented the budget request for the Canadian River Project to Congress shortly before leaving office in January 1961. As the Lubbock newspaper put it, "This [action] apparently generated the impression that the money is practically assured. . . . Such optimism is not justified. The Presidential recommendation will prove meaningless unless Congress adopts it." Earlier attempts at obtaining the funds had died in committee due to the lack of water-purchase contracts, among other causes. On one occasion Senator Wayne L. Morse of Oregon had blocked appropriations for the dam because of his opposition to federal support for such projects as a matter of principle. That June, Representative Walter E. Rogers from Pampa assured A. A. Meredith that he would "work hard" for a $4.2 million appropriation of the Bureau. Rogers pointed out, "[P]rotecting the project was no simple task under the handicaps of not being in the position to seek and press for construction funds." Rogers may have been thinking of Senator Morse's efforts when he reminded Meredith that the hiatus of the previous decade had led to multiple attempts to cancel the project, which "has been watched with a jealous eye," and forcing Rogers, George Mahon, and other Texas solons to fight off the project-killers.[2]

Back in Texas, the Bureau began obtaining land for the project. Although the ranchers seem to have been willing to sell, all was not sweetness and harmony. Prime irrigated pasture, grazing land, and water sources would be drowned or buried by the reservoir and dam. A man named Wilbur Rogers called George Mahon in early September 1961 and warned that there were problems. "Chester Gray [*sic*], a lawyer with the Bureau at Amarillo has offended some of the land owners by saying [that the Bureau] would use all the land destroyed et cetera for recreation et cetera." Twenty years before, the owners of the Bell Ranch had raised similar concerns over recreation on Conchas Reservoir, arguing that their private property and riparian rights to the river (and lake) banks overrode the public's right to picnic, sun, and fish from those banks that the ranch still owned. New Mexico argued in turn that all waters belonged to the state because of the territorial and state constitutions, and while the ranch might own the banks and even the riverbed, the waters remained open for all members of the general public to use. The New Mexico Supreme Court had sided with the state in a 3–2 decision. The dissenters expressed concern that common law of riparian and private property remained in effect and should not have been struck down, particularly those aspects related to private property rights. Fortunately for proponents of the Canadian Project in Texas, landowners were more amenable, especially those along the pipeline right-of-way. They received restitution for any crops lost to the construction and free access to the project's water if they so chose.[3]

Despite problems in Washington and among the affected landowners, on June 30, 1962, dynamite blasts echoed off the steep valley walls and the first earth moving equipment set to work on the enormous project. The most visible section of the project would be what the *Amarillo Times* had irreverently if accurately termed a "big dirt stopper." Technically, Sanford Dam came under the heading of an earthen gravity dam with concrete spillways and works, topped by a road. In layman's terms, the massive stone and dirt structure held back water by the sheer mass blocking the Canadian's channel. A concrete "morning glory" flood-outlet, an emergency exit for water that superficially resembled a bathtub drain, and concrete spillway allowed water to continue downstream and protected the dam should a large flood come down the river. Concrete also formed part of the heart of the dam after workers had dug twin trenches into the river's sandy bed to a depth of 30 feet (9 m) and 100 feet (30.5 m) across in an attempt to find a solid footing—or at least to create a firm enough and deep enough foundation that water could not leak out from under the dam. Above the hardened riverbed, workers used bulldozers and other earthmovers to pile up rocks, pack dirt around the rocks, and then work more large rocks into the upstream side of the dam to armor it from the water and waves. Adding to the problems for contractor H. B. Zachary, workers found more vertical tunnels, called chimneys, in the canyon walls and floor than had originally been anticipated. These had to be filled or stabilized to prevent dam material from slumping into the voids and necessitated redesigning part of the dam and its works. On the north end of the dam workers poured and shaped the spillway, a concrete trough that led into a stilling pool where water from the lake could settle before flowing downstream toward Oklahoma. This "stilling" helped keep the river from eroding that part of the dam and the downstream riverbed. On the northern wall of the canyon, workers added a form of cement called grout, pumping it into cracks in the rock walls to make them more resistant to water erosion.[4]

The final structure loomed large indeed. It rose 228 feet above the floor of the valley. The dam's crest extended over a mile from north to south at 6,380 feet from end to end, topped by a two-lane state highway. Forty feet wide at the crest, the massive rolled-earth structure covered 1,900 feet (580m) of the river valley. To make the structure, workers shaped 15,308,000 cubic yards of dirt, adding 347,650 cubic yards of rock riprap on the outside of the dam. A further 448,170 cubic yards of rock armored the channels through and below the dam, protecting them from erosion. Most of the stone came from quarries near the dam site, which helped reduce the final cost of the project slightly. Not elegant or graceful like Hoover and Glen Canyon Dams on the Colorado River, Sanford Dam relied on mass of materials to keep the river contained. Although large, this style of dam was relatively easy to construct, and Zachary Company faced no major difficulties during the five-year building process.[5]

Project supporters and potential contractors knew that laying the aqueduct would be more difficult than building the dam. The pipelines required six pumps, a number of surge tanks, and two reservoirs, one in Amarillo and one in Lubbock. According to the Bureau's plans, two main runs of concrete pipe extended from the pumping station at the dam—a shorter one east to Borger and Pampa, and the 325-mile (545 km) run south as far as Lamesa. For much of the route the pipeline would follow a state highway, assuring project managers and contractors of ease of access but complicating right-of-way considerations. Rather than condemning land outright or trying to buy all the acreage involved, the Bureau and CRMWA reached compromises with landowners and agreed to pay for crops and pastures lost or damaged during the construction process. Obtaining rights-of-way from the Texas Department of Transportation for the section of pipeline that passed under a highway in Lubbock generated the most paperwork of any section because of liability questions: If road vibrations damaged the pipeline, who had to pay for repairs? If a leak from the damaged pipeline caused highway problems, where would the money come from to fix it? The process took time, but eventually all rights-of-way became available for the Bureau's enormous project. Had the pipeline been built when originally approved, it would have been the longest pipeline aqueduct in the world, but by the mid-1960s projects in California and elsewhere on the globe outdistanced it.[6]

As with many other federal projects, debate and discussion over funding the project continued after dirt began flying. This time the dispute came about due to fish and ducks. Of the proposed $96,090,000 cost of the Canadian River Project, $3,000,000 of flood-control protection for Oklahoma and Texas could be deducted from the final bill to the Canadian River Municipal Water Authority. No one disputed that part, but the additional deduction of $1,612,000 for fish, wildlife, and recreation raised hackles on Capitol Hill. In September 1963 Floyd Dominy and Representative Joe D. Ewing of Tennessee exchanged heated words during a meeting of the House Appropriations Subcommittee for Public Works. The Tennessee Democrat wanted to know why the federal government should pay for fishing and recreation that benefited only local businesses and lake users. Joe R. Pilion, a Republican from New York, complained further that the project was "scraping the bottom of the barrel" in terms of cost/benefit because it would provide only $1.40 in benefits for every dollar spent on it. The wildlife discount remained in the package despite the representatives' objections, leaving CRMWA with "only" $92,960,000 to repay, beginning whenever the first water reached the first member cities.[7]

Even as politicians in Washington, D.C., cussed and discussed funding different aspects of the Canadian River Project, two fighters for the dam retired from the fray. John McCarty had left the world of journalism and turned to insurance, real estate, and painting, as well as compiling the first history of Old Tascosa, the wild and wooly cow town that had lived and died beside the Canadian in Oldham County. McCarty remained interested in the Canadian River Project but no longer worked on conservation efforts, in part because of a serious and lingering depression following the death of his beloved wife, Susan, in 1957. McCarty also faced major financial setbacks beginning in that same year that led to his retiring further from public life. A. A. Meredith, former manager of the city of Borger, Rotarian, CRMWA secretary, and tireless campaigner for the project, had grown no younger during the decade prior to construction of the dam. He fell ill in early 1963 and passed away on April 13 of that year. He had received many awards over the years, including "Man of the Year" from Borger's Kiwanis on two occasions. In 1962 the Department of the Interior, via the Bureau of Reclamation, presented Meredith with the Conservation Service Award, an award given "for furthering of the cause of national resource development [that] is the highest honor that

Figure 7: Dirt work in progress at Sanford Dam site. Used with permission of the Hutchinson County Museum, Borger, Tex.

the Department of Interior can confer on a private citizen." As a gesture of respect for the man who had done so much to make the dam and aqueduct a reality, Congress voted to name the reservoir in his memory: Lake Meredith. It was fitting to name one of the last great Progressive Era conservation projects after "Double A."[8]

Although built long after the Progressive Era in American politics had passed, Lake Meredith could be seen as part of that era. Meredith's and McCarty's public actions reflected the same philosophy that animated the efforts of John Wesley Powell and Gifford Pinchot. Conservation as these men viewed it entailed the best use of natural resources for the most people over the longest period of time. As Powell had urged westerners to do, the residents of the South Plains and Panhandle had worked together to form a water district so that the "wasted" waters of the free-flowing Canadian could serve almost a million people who otherwise might run short. The federal government provided some of the resources to make this possible, but the Texans had done most of the work and would pay back most of the cost.[9]

By putting a dam and lake on the Canadian, members of the general public provided for more than just drinking water. No longer would the people of Oklahoma face the prospect of untamed floodwaters destroying their infrastructure and livelihoods. Upstream of the dam, the public lands surrounding the lake beckoned those interested in new recreational opportunities based on the Canadian, including swimming, fishing, boating, water-skiing, hunting, and exploring the Alibates flint quarry archaeological site. In a letter to President John F. Kennedy, Meredith had explained that the new dam and lake were "of utmost importance of the much needed improvement of family life throughout the United States." He continued in the next paragraph that people from the Panhandle had to travel great distances in order to find opportunities for the recreation that would "seal family ties that will endure for life," and presumably improve society. As seen by those interested in conserving the Canadian through wise use, Lake Meredith served a high calling indeed.[10]

However, fewer people than before agreed that putting "big dirt stoppers" in rivers was the best way to conserve, or better, to preserve rivers and their valleys. In many ways John McCarty and A. A. Meredith serve as examples of the last of the traditional conservationists. The generation that followed them viewed such certainty with suspicion and wondered if turning rivers into nothing more than chains of carefully managed lakes was truly the best and wisest use of the land and water. More and more people began expressing their doubts.

The way increasingly large numbers of Americans thought of their surroundings had changed since World War II due in part to rising affluence and the spread of suburbs. As historians Samuel P. Hays and Adam Rome point out, there had always been individuals who valued preserving parts of the landscape and the resident flora and fauna over development, for whom "wise use" included observation and recreation but not exploitation. These people, such as naturalists John Muir, Aldo Leopold, and Bob Marshall, began as a minority within the larger conservation movement. After the Second World War that ratio began changing as more people moved out of the cities and earned more disposable income. Families bought houses in the new suburbs in order to escape urban crowding and to provide their children with cleaner air, lawns, gardens, and more access to "nature." At the same time, growing numbers of people took up outdoor recreation, visiting national and state parks and forests, getting fishing and hunting licenses, buying camping equipment and trailers. People who enjoyed paddling a canoe through the rapids of the Colorado River in Arizona, to use a well-known example, did not necessarily want to see their river turned into a lake. And suburban homeowners became increasingly concerned about the fate of the birds that had once sung from the trees now being cleared in order to make room for the next suburb. Conservation did not seem to be enough. Perhaps time had come to set aside land and water, to preserve them as they currently were, for the benefit of present and future generations.[11]

Changes were also afoot in the national economy, although their effects on peoples' approach to the environment remained less obvious in daily life. Prior to 1960, manufacturing dominated the United States, with industrial production and agriculture two of the largest sectors of the total economy. After 1960, improved technology and the export of some jobs led to the beginnings of "deindustrialization." Technology and service-sector work gradually replaced heavy industry in economic importance. In part because of this decreasing economic clout, regulating bodies and government agencies became less willing to ignore air and water pollution for the sake of local incomes and wealth. The highly polluted Cuyahoga River in Cleveland, Ohio, accidentally set alight yet again in 1969, became a national symbol of industrial pollution and further stimulated environmental protection activists. In the face of national media attention, shutting down or more tightly regulating plants that no longer contributed to the local economy became less onerous. Hays described the shift in what people wanted in their environment as being from beauty—attractive parks—to health—reduce or eliminate dangerous emissions, leaks, and products—and finally to permanence. Permanence marked where the new postindustrial conservation movement differed the most from the earlier variety.[12]

Conserving no longer sufficed. The time had come for preserving the environment. Rivers formed one of the battlegrounds for the new generation of conservationists, and the question of "wise use" versus "preservation" loomed large, notably in battle over dams on the Colorado River and the new concept of "in-stream flow." Water law in the states generally west of the 100th meridian focused on "beneficial use" of the streams and rivers' waters. Domestic, agricultural, and eventually industrial uses fell under this category because the water was "doing its duty" for people and their animals and crops. If any water remained within the riverbanks after all the users had been supplied it was a happy accident, and for some rivers a rare occurrence. Fish, reeds, beaver, and crayfish lacked water rights even though they were the truly senior users of streams. Fishermen, boaters, and swimmers also lacked rights even to have access to some streams, depending on state laws concerning who owned the banks and bottoms of navigable waterways. Although more and more states and the federal government set aside land for parks where people could enjoy the uplifting and improving presence of "nature," western water law failed to consider recreation and wildlife preservation. The author Charles F. Wilkinson has argued that "the lords of yesterday" govern this economic-use-based approach to resources; that the philosophical views of late-nineteenth-century miners, lumber magnates, and irrigation developers still dominate. As a result, through the early 1950s the effects of dams on recreation and wildlife remained at best tertiary considerations for those in charge of building and maintaining the huge structures, at least until David Brower and the Sierra Club brought the matter to the attention of much of the nation.[13]

The Colorado River in Arizona became a focal point in the battle between river developers and river preservationists. It had been an infamous stream since Anglo Americans first set eyes on the warm, muddy, unpredictable, often boulder-choked river, one of the largest in the West. Until John Wesley Powell led his expedition through its depths the Grand Canyon portion of the river had remained "the great Unknown," one of the last unexplored parts of the continental United States. Farther downstream, tapping the wild river's flow proved tempting and nearly impossible well into the first quarter of the twentieth century. First in California and then at Boulder, Nevada, dams caught the river's waters, holding them back for the later use of thirsty plants and people. Boulder (later renamed Hoover) Dam also generated electric power for rapidly growing western cities. By 1960, a comprehensive plan of river management drawn up by the Bureau of Reclamation envisioned dams on the main stem of the Colorado at Echo Park in Dinosaur National Monument, Glen Canyon, and within the Grand Canyon, among other locations. But a growing and vocal number of Americans wanted a free-running river more than they wanted another power-generation lake, especially if that lake compromised the beauty and the recreational opportunities afforded by a national park. The story of the battles between the Bureau and preservationists has been told elsewhere, but suffice to say that the 1950s and especially the 1960s marked the beginning of the end of the big-dam era in the American West.[14]

It is worth noting that people outside of the circle of wilderness proponents expressed distrust in further giant water projects and the continuing calls of "drought, drought!" Arthur Carhart of the Department of the Interior penned an article in 1950 entitled "Turn Off That Faucet!" in the *Atlantic* arguing that conservation and preservation were synonymous and necessary. He pointed to water shortages caused by wastage, to inefficient land and timber use in Iowa that led to water problems (something J. W. Powell had also expressed concerns about), and urged an end to pollution that not only ruined surface water but forced people to turn to nonrenewable groundwater. Drought in the Northeast in the 1960s inspired more talk of dams, bringing author and westerner Wallace Stegner into the fray with an article in the *Saturday Review* pointing out the "Myths of the Western Dam." "The questionable dams are never simple water holes," Stegner explained, highlighting the power dams (Grand Coulee), politics, and the problem of self-perpetuating bureaucracies, especially the Bureau of Reclamation and Army Corps of Engineers. "Bureaus loaded with know-how are short on know-whether, and competition, boondoggling and the pork barrel hardly substitute for intelligent examination of the public interest." He supported Utah senator Frank Moss's proposal to combine all resource agencies—mines, water, navigation, forestry, soil—into a Department of Natural Resources.[15]

A longer article in the April 1965 issue of *Fortune* focused on the municipal and cultural aspects of "water shortage." William Bowen outlined the reasons for recent and historic droughts and dry weather, and then turned his attention to water use and policy. "Water shortage is a frame of mind," he told the businessmen reading his article. The problem stemmed from the cheapness of water. Water cost too many people too little. Individuals and unmetered businesses had no desire to conserve because they did not know how much the resource actually cost and therefore did not care how much they wasted. Those who had to pay for their water found increasingly efficient ways to use less and less in their manufacturing processes. Bowen examined the costs and necessity of desalination plants to convert seawater into fresh and then focused on the nation's greatest water user—irrigated agriculture. Bowen considered many of the irrigation uses, including growing cotton in Arizona, to be foolish when compared to the return for industrial use of water. He cited a University of New Mexico study that found an acre-foot of water for irrigation in the Rio Grande basin added fifty dollars to the gross domestic product, while an acre-foot used for industry contributed three thousand dollars. Evaporative waste and water pollution also raised his ire, and he concluded with the example of the German Ruhr, where limited water supplies remained clean and usable because of how the West German government charged water users. The more polluted their discharge, the higher the fees that factories and municipalities paid. The government then used those fees for cleaning the water. This encouraged businesses to do their own purification and resulted in efficient use and cleaner water for all users. In short, the only shortage was of realistic billing and efficient use and management, not of total water supplies. Bowen ignored recreational use of water and the ongoing preservation versus conservation debate in his article, but one suspects that the David Browers of the world agreed with Bowen's arguments, to a point.[16]

Back in West Texas, preservationists seem to have ignored Sanford Dam and Lake Meredith. The region tended to be socially conservative, and larger national movements such as preservation seeped in rather more slowly than they did into Austin or San Francisco, for example. There may simply have been no one interested in preserving the Canadian River in its free-flowing state. Unlike structures outside of Texas, the Canadian River dam and lake occupied formerly private land and provided those members of the general public interested in seeing the Breaks with their only opportunity to hike, boat, fish, or hunt along the river. This access had not been available before, and great excitement greeted the news when President Lyndon B. Johnson signed a bill authorizing the Fish and Wildlife Division to run the shores of the lake and oversee fishing and recreation in cooperation with the Canadian River Municipal Water Authority (CRMWA). The recreation area also gave the public access to the Alibates flint quarry, an early American Indian site on the south side of the river. This was a new commons, much larger and easier to access with more amenities than the riverbed had. Panhandle residents interested in keeping the Canadian free-flowing may also have looked at the decades of strong support for the project and decided that there were better battles to fight: protecting Palo Duro Canyon, Caprock Canyons, the various refuges and grasslands. The construction of another dam upstream of Sanford may also have convinced preservationists that protesting the dam would be a lost cause.[17]

As Zachary Company's subcontractors piled dirt upon rocks and poured concrete in Texas, a similar scene unfolded in New Mexico. The Canadian River Compact granted New Mexico 200,000 acre-feet of storage downstream of Conchas Dam, and the state legislature had set aside funds for that second structure. In 1959 engineers considered and surveyed several sites, including the Ute Creek–Canadian confluence and farther downstream at the Dunes site. Dunes, below both Ute and

Tucumcari (also called Revuelto) Creeks, would catch more water, but the Ute location provided better rock for a dam. Dunes took its name from the sand dunes stretching north from the Canadian's northern banks, something that dam builders preferred not to have to deal with. Ute Creek provided more water than Revuelto Creek did, and here the Canadian flowed through a relatively narrow canyon with only fifty-seven feet of sand and silt between the water and the bedrock. The state water engineer approved the Ute Creek location, and by late 1963 construction finished on Ute Dam. The final structure, an earthen dam like Conchas and Sanford, stood 148 feet (45 m) high and 2,050 feet (625 m) along the crest. Two years later, its downstream sister also began catching water.[18]

Sanford Dam did not require the time and effort needed for the construction the aqueduct. Pumps lifted water from the reservoir up to the first division point. From there the water would flow east to Borger and Pampa, or south. More pumps would keep the vital fluid moving to Amarillo, where it "rested" in a reservoir for about thirty-six hours before moving south along a large underground pipeline. Once workers made all the necessary connections and after Plainview took its share, the water would stop again at Lubbock. There a central water treatment plant purified and chlorinated it before the water was distributed to the other towns as far away as Lamesa.[19]

J. R. Fulton Company, an oilfield pipeline specialty company belonging to the owner of the Alamositas Ranch on the Canadian, won the contract for laying the pipeline. This involved digging trenches for the pipe, some of which at ninety-six inches (244 cm) in diameter were large enough to drive a small car through. For the main run of the aqueduct, workers excavated a trench ten feet across and up to twenty feet deep. Then they laid the concrete pipes on a pad of sand, aligned them, and joined the sections together with special joint material. This process involved a great deal of very difficult and precise work by the excavation crews and the men literally in the trenches guiding the pipe into place because the pipe had to slope a certain amount over the length of the 300-mile (483 km) run. One of Fulton's employees and a Bureau engineer together devised a faster and equally precise way to dig the trenches, in the process inventing a new type of machinery to do so. Once a shallower trench had been dug, the new machine cut a smoother and more precise channel. As a result, the more precise fit of the pipes required less time for final adjustments and also made backfilling the trench easier. This, plus a series of mild winters, allowed work to move over twice as fast as planned, and the final aqueduct finished almost $11 million under budget as well as ahead of schedule. The pipeline also changed the water equation of the Canadian River in a way that most people did not consider.[20]

The Canadian River Project entailed an interbasin water transfer. Once the taps in Plainview, Lubbock, and Lamesa opened, the river's waters entered the Red River, Brazos, and Colorado watersheds, albeit frequently after having been used for irrigation following wastewater treatment by the various cities and towns. Water used in Borger and Pampa returned to the Canadian's main flow, while Amarillo's treated water went in part north, back to the Canadian via East Amarillo Creek, and partly to the south, into the Red River watershed. Plainview's used water also flowed into the Red River. Lubbock's discharge flowed into a fork of the Brazos River, and that from Lamesa eventually wound its way into the Colorado River. Unlike the case with other interbasin transfers in the West, the only eyebrows this one raised came from the state water board, over how Lamesa disposed of its water—via irrigation or plain outflow. No one seems to have batted an eye at redirecting two-thirds of the river's projected flow into other watersheds. If asked, those citizens of Texas and Oklahoma downstream of Sanford who remembered the flood of 1941 probably considered the redirection a good thing. Those South Plains residents now guaranteed a supply of drinking water expressed no concerns about in-stream flow below the dam or interbasin transfers. They were delighted just to turn the tap and have drinkable water coming out the faucet, and in Tahoka's case water that tasted better. The arrival of Canadian River water was a happy day for the people of Lamesa, Tahoka, Lubbock, and other southern cities, generally undisturbed by environmental preservation concerns or by the rumbles emerging from Amarillo's water users.[21]

A loud "pa-THOOY! Blech!" greeted the arrival of Lake Meredith water to Amarillo. The unique flavor of Canadian River water should have come as no surprise to area residents, who had been warned numerous times that it contained more salts than did the well water they had used for seven decades. Perhaps Amarillo's water users thought that the water would be flavored more like fancy *eau mineral* or taste slightly different but still crisp. Instead, when the salt-flavored fluid poured out of their faucets, Amarilloans protested loudly and often. Some claimed that the water would cause medical problems, while others insisted that it was destroying their water heaters and pipes, benefiting only the plumbers now flocking to the city. A petition drive for a return to well water ensued. Amarillo's city commission repeatedly assured residents that the water was safe and reminded taxpayers that the city was obligated to pay for 35,000 acre-feet/year even if it returned to pure well water. In March 1969, nine months after the first Lake Meredith water emerged from the taps, Amarillo residents voted in a referendum. Did they want to go back to pure well water and still pay for the unused lake water, to go half and half, or to continue as things were? The majority opted for a blend of lake and well water. People continued to grumble and muttered about the deposits appearing in their pipes and the early deaths of their water-heater tanks, but Amarillo took its share of the river's apparent bounty.[22]

Two years before the first lake water reached Amarillo, area notables had gathered once more in the Breaks, this time to celebrate completion of Sanford Dam. Only major floods in the spring of 1965 delayed the work; H. B. Zachary had pulled its crews from the Canadian River Project for a few weeks in order to make emergency bridge and road repairs in Colorado and Kansas. Despite the interruption, Sanford Dam stood finished and solid, a testament to the persistence and foresight of the people of the Texas Panhandle and South Plains. Governor John Connally, Representatives George Mahon and Walter Rogers, and Interior Secretary Stewart Udall joined Mrs. A. A. Meredith and other local notables on an unpleasantly chilly November morning in 1966. This time, the preacher from the First Christian Church of Pampa offered thanks. Not everyone overflowed with gratitude to the Bureau, and Representative Rogers wrote George Mahon that he did not listen to the talk because of the severe cold. He was also irked that Senator Ralph Yarborough had been left out. "It presently appears that full credit will be taken by an erstwhile columnist in Amarillo; several citizens of Borger who had very little to do with it, if any; a Republican senator; and several miscellaneous Republican bad mouths who called it socialism until it was completed, but who now claim to be mother and father of it," Rogers snarled. In all likelihood most of the people celebrating the completion of Sanford Dam worried less about the credit (or blame) for the final results and just wanted their water and a new place to play in it.[23]

Far below the people gathered on Vista Point, water shimmered and reflections danced on the water as the wind blew across the new lake. Those interested in hunting looked forward to the opportunity to stalk deer and wild turkey, to lie in wait for the geese and ducks drawn to the new lake. Boats already nestled against the new docks and marina, and wardens would soon release the first game fish into the lake. If anyone regretted the loss of the channel catfish, Arkansas shiner, and other piscatorial natives, they kept silent. Lake Meredith stood for security in an

Figure 8: The big dirt stopper: Sanford Dam and Lake Meredith. Used with permission of the Hutchinson County Museum, Borger, Tex.

increasingly uncertain part of the country. On November 1, 1966, watchers probably felt that for today, the changes rippling through their communities and nation could be ignored. Here was stability and progress in the form of rising water and a massive dam.[24]

In many ways, the period between 1960 and 1969 was a lull in the hydrologic uncertainty that was life in the newly enlarged Canadian River watershed. Marked by neither terrible drought nor record floods, the river flowed until stopped first at Ute Dam and then at Sanford. Harmony, more or less, reigned among the eleven cities of CRMWA, which ceremoniously handed over the first payment check to the Bureau of Reclamation in 1969. In some ways the watershed had grown, now stretching far to the south almost to the southern tip of the Llano Estacado. But as 1968 turned into 1969, trends and patterns that had been hidden or ignored became evident. The next decade marked a return to the uncertainty of the 1950s as infighting within CRMWA, dropping water tables, and rising federal influence in the region all dimmed the enthusiasm and optimism of those dependent on the Canadian's waters.

RED WATER, BLACK GOLD

Chapter 8

Changes in the River: Lawsuits, Leaks, and Brine

The office of John Williams, the general manager of the Canadian River Municipal Water Authority, provided a lovely view of the dam and lake. By all rights, since the Bureau had finished building the dam, the aqueduct had been completed and delivered water without incident, and the lake grew deeper and larger during the early 1970s, Williams's days should have been as tranquil as the scene outside his window. But it was not to be. Amarillo and Lubbock fought over paying for the running of the dam, a tornado carried off part of the riprap armoring the upstream face of the dam, and by 1980 Lake Meredith's waters had taken on a very distinctive salty flavor thanks to artesian brine springs upstream. And Williams had to mediate, placate, repair, and manage the entire water system and the people who depended on it. At least his office and residence provided some of the best views in the Panhandle. Floor-to-ceiling windows in the new CRMWA headquarters building, upslope from the dam, looked north into the swelling, windswept lake. Small boats cruised back and forth, and hawks, gulls, and the occasional bald eagle soared past the building, riding the canyon's winds. The directors beheld a very well conserved and managed scene from CRMWA's boardroom window wall. The papers on the boardroom table told a different story.

The 1970s on the Canadian River could be summed up as a decade of lawsuits, leakage, and brine. Two major suits, one between Amarillo and Lubbock and another between a contractor and the Bureau of Reclamation, kept lawyers employed and CRMWA board meetings interesting. Threats of litigation over water contamination from an as-yet-unbuilt sewage treatment plant also required attention, lest CRMWA find itself involved in a federal water-quality lawsuit. Seepage from the Ogallala

Aquifer, previously an important source of flow for the Canadian in Texas, began declining as a consequence of rapidly increasing groundwater irrigation in the area. And the lake's waters turned increasingly brackish and briny, sending experts from the Bureau of Reclamation on a hunt for the source of the salt and then for possible solutions to the problem.

The lawsuit that consumed much time and patience during the first half of the 1970s stemmed from a number of sources, including one not at all related to water or infrastructure. During the early 1960s the Department of Defense underwent a periodic reconsideration of its training and staffing needs. The base at Amarillo duplicated the functions of Reese AFB in Lubbock, Tinker in Oklahoma City, and Canon AFB in Clovis, New Mexico, and the military no longer needed it. As a result, in 1964 the Air Force announced that it intended to close the Amarillo Air Force Base, one of more than 500 facilities to be shut down or consolidated.[1]

The Air Force had deactivated Amarillo's WWII training base in 1946 and then reactivated it in 1951 during the Korean Conflict. Relations between locals and the military personnel had generally been cordial as well as profitable. For example, the *Amarillo Globe-News* and local leaders named base commander General Dwight O. Monteith Amarillo's "Man of the Year" in 1963, in part for the landscaping and other improvements he made to the base. After spending funds for trees to plant along base roads, General Monteith asked Amarillo residents for their excess ornamental and flowering plants to use in other parts of the base landscaping. The general sent airmen out to collect the donations, and the men also tidied up the yards whence their new landscaping came, making both the city and the air base more attractive.[2]

However, good relations meant little in the big picture of national defense needs. In November 1964 Secretary of Defense Robert McNamara announced the pending closure of Amarillo AFB and scores of other military installations. As had been feared by city leaders, when base-closing day came in July 1968, Amarillo's economy stumbled as real estate prices sank due to lack of demand, school enrollment decreased, and businesses that had catered to Air Force personnel shut their doors. In fact, Amarillo's overall population declined from 137,969 in 1960 to 127,010 in 1970. Amarillo's rapid growth during the first half of the decade masks the true effect of the closure: roughly 40,000 people packed up and departed Amarillo between 1967 and 1970. This near-disaster put the city's tax rolls into a tailspin, and the resulting mood in city hall very much resembled that of the money-starved late 1940s.[3]

As Amarillo scrambled to recover lost people and revenue, Lubbock's representative to the CRMWA board announced that a new payment plan was needed, one that would hit Amarillo's finances even harder than the current payment system did. The original payment plans that area residents read about in their newspapers based everything on the costs of water, the dam, and pipeline, omitting the fixed costs and day-to-day maintenance expenses for the project: employee salaries and benefits, phone lines, heating and lighting CRMWA offices, some chemical costs, and other such things. In March 1968, when the first bills came due as water taps turned on, it had been decided by the board that "each of the cities would be charged with one-ninth of the total fixed cost each month plus a variable charge based on each city's estimate of its power requirements for that month." In other words, the board divided *everything* equally among the eleven cities and towns, then added in the specific charges for the water they used. Come December, the board further clarified this formula to state that each city would get a three-part bill: one-twelfth of the annual fixed costs (administration, some chemicals, the dam), the cost for the water delivered to the individual city, and the construction repayment cost. The fixed costs included a per-mile of aqueduct charge so that each town paid for its share of the pipeline.[4]

J. Ray Dickey of Lubbock, a former city council member and owner of a major regional car dealership, protested this method of billing. It appeared that Lubbock would be paying more than its fair share of the operation and maintenance (O&M) costs. CRMWA manager John Williams replied that while that might be so, it was the formula to which everyone had agreed in 1960. Williams also pointed out that if Lubbock was paying ten thousand dollars more than Amarillo per year, that worked out to "one cent per thousand gallons overall difference over the 50-year period" of paying down the cost of the dam and aqueduct. And it was too late to change next year's billings. Dickey took this news back to Lubbock and after six months of consideration, presented a new formula and plan in July 1969. Dickey argued further that the small cities were hardest hit by the Bureau's billing plan, since they were at the far end of the pipeline and used the least water. Amarillo's representatives disagreed with his assessment.[5]

Matters came to a head at the April and July 1970 CRMWA board meetings. Dickey stated succinctly, "Lubbock is not happy about paying forty-seven percent of the cost of fixed operation and maintenance at the dam." Lubbock would pay a pro rata share and everyone else should too. In 1968 the O&M costs had been almost $300,000, of which Amarillo and Lubbock each should have paid $111,174, their share. The 1960 agreement would have Amarillo paying pro rata for aqueduct and O&M and 162 percent of its "share" of the dam and reservoir costs, while Lubbock paid 42.5 percent of the dam and reservoir costs. That was not what was going on, in Lubbock's view, and they wanted to change things. CRMWA president George Finger, the board member from Borger, reminded Dickey that they all had to abide by the contract. Dickey replied, "Lubbock does not want to get locked in on 48 percent of the fixed O&M costs." After more discussion, Finger reminded everyone that a change in billing required a unanimous vote, something the proposal lacked. Lon Hill of the Bureau of Reclamation stated that the Bureau did not care how it got paid, so long as it did get paid. When put to a vote, Amarillo, Borger, and Pampa opposed the proposed change, with the southern towns voting for it or abstaining. On a more positive note, the meeting closed with the cheerful news that, according to National Park Service personnel, observers had counted a record 350,000 ducks on the lake that winter.[6]

But the debates over billing continued, growing more heated. The July meeting focused on the budgeting question. John Williams and the Bureau remained neutral on the matter. J. Ray Dickey (Lubbock) and George Finger (Borger), A. F. Madison (former Amarillo mayor and current dairy manager), and Al Banasik of Amarillo went around and around as other board members commented on the proposals and arguments. Dickey reminded everyone that "the city of Lubbock is not at anybody's throat and that all that the City of Lubbock wants is what is fair and equitable." The current system seemed unfair to Lubbock and the southern towns, while Lubbock's proposed plan hit Amarillo's, Borger's, and Pampa's wallets. Amarillo protested that the vote to change the billing system had not been unanimous, and therefore it did not change the billing plans. Lubbock disagreed, as did Brownfield, Plainview, Slaton, Tahoka, Levelland, Lamesa, and O'Donnell. Officially and for the public record, no suit had been filed and there were no threats of a lawsuit pending, according to one of the few points of agreement among the board members. In reality, both sides began planning for a court confrontation unless someone could find a compromise.[7]

Pretty much everyone on the board sensed a lawsuit in the offing by October 1970. Another vote on the Lubbock pro rata proposal yielded nine in favor, four opposed, and two abstaining, but the revised payment plan remained illegal because it violated the 1960 contracts and had not been passed by all the cities even though a majority of the board of directors supported the change. Apparently feelings had been hurt and feathers ruffled, because J. Ray Dickey admitted that "perhaps he got overzealous," but he assured everyone that it was nothing personal and he only wanted what was "fair and equitable." Although they disagreed, Amarillo's leaders could not fault Lubbock for wanting to pay less, especially after a tornado ripped through Lubbock on May 11, 1970. The storm tore up a quarter of Lubbock, including the downtown, killed twenty-six and injured more than two hundred people, and damaged the water pipeline and pumping station. It was just that Amarillo, along with Borger and Pampa, did not care to pay any *more* than what they considered to be "fair and equitable." In April 1971 Amarillo sued CRMWA in order to enforce the original payment plan agreed to in 1960.[8]

The court battle may have come as a relief to the rest of the board of directors and member cities. With the dispute now outside of the CRMWA boardroom, the directors could turn their attention to other problems. For example, water-quality concerns took up a great deal of the board's time. The city of Dalhart still dumped its waste into Rita Blanca Creek, one of the main tributaries of the Canadian River upstream of Lake Meredith. This practice followed the old and somewhat correct argument that "dilution is the solution for pollution," which assumed that with the addition of enough cleaner water and the decay of the organic pollutants, dilution reduced the health hazards of chemicals and sewage. However, because now people were drinking from the Canadian and fecal coliform bacteria counts now had to be reported to the newly created Environmental Protection Agency along with the Texas State Department of Health, acceptable practices no longer included raw dumping—thus the $2 million included within the Canadian River Project to move and renovate Amarillo's sewage treatment plant that had been upstream of Lake Meredith. CRMWA watched but did not threaten Dalhart in 1969 because the city was trying to find solutions while raising funds for a new water treatment facility. Dalhart's efforts including letting a farmer irrigate with the waste (a form of liquid fertilizer, sort of) until it got the new treatment plant constructed. The Authority also kept a wary eye on feedlots in the area for the same reason.[9]

Another water-quality concern, and one that was not as amenable to solution, came from a combination of geology, climate, Ute Dam, and irrigation. As mentioned before, the Canadian River had a reputation for poor water quality. Prehistoric oceans had contributed to the modern river's saltiness by laying down thick layers of gypsum, anhydrite, and salt as the oceans advanced and retreated over millions of years. Eventually, the river and its tributaries eroded their way down into or close to these layers. Groundwater also dissolved the salts and brought them to the surface.

Saline springs existed near the Texas–New Mexico border, and Hispano pioneers from the plaza of Las Salinas had used the thick brine to make salt both for personal use and for sale to neighboring ranchers. Hispanos, Anglos, and others generally regarded the Canadian's brackish, muddy waters as not worth drinking if there were other options available.[10]

As the river's waters backed up behind Sanford Dam and evaporated, the salt in the water became more concentrated and more apparent. At the January 12, 1970, board meeting, W. B. Hoover of Amarillo complained that gas wells under the rising water leaked natural gas and salt into the lake, affecting the water's quality and taste. John Williams reminded everyone that the salt predated the lake, adding that recent inflows and the lake's fall turnover had improved the water quality noticeably. The river at low flow carried 1,000 parts per million (ppm) of chloride, which the three years of evaporation between 1965 and 1968 had not helped. The wells in question, Williams explained, had been capped or relocated. Hoover asked about the dry holes that predated the 1930 Texas Railroad Commission capping regulations. Williams said the board would look into the question, but either they were not a problem or they had already been dealt with because no further evidence of contamination concerns from the old wells appeared in the CRMWA minutes.[11]

Investigators for CRMWA and the Bureau of Reclamation found one possible major salt source upstream of the lake. Artesian brine springs had appeared in the river valley below Ute Lake. These seeps postdated Ute Dam's construction and produced very salty water that trickled into the already saline Canadian. In January 1971 the board discussed allocating funds toward further investigation of these springs. The Bureau also asked Congress for $25,000 in order to study the problem. Over the winter a contractor looked at the water in Ute, the water coming from Ute Lake, and the springs and concluded that yes, the Lake Meredith salt problem stemmed from the new salt springs downstream of Ute Dam.[12]

What had happened? Geological investigations in the area eventually provided a possible answer. As the water backed up behind Ute Dam, the weight of the water forced some of the fluid down through the old riverbed sediments and into the underlying salt- and gypsum-rich rocks. The subsurface strata in this area dipped slightly to the east, providing a conduit and direction for the water to travel along and between the layers of more waterproof stone. As it moved, the water dissolved some of the salt, bringing it to the surface where the water emerged from between the rock layers downstream of the dam. The State of New Mexico no doubt took a dim view of the simplest solution to this problem: dewatering Ute

Lake until the brine seeps stopped flowing. As a result, CRMWA and the Bureau never seriously discussed removing or lowering Ute Dam. So what were the Authority's options?[13]

CRMWA initially took a "watch and wait" approach. In April 1973 the lake continued to rise, reaching what would prove to be its all-time record high level of 101.85 feet (31 m) as a wet spring sent water pouring into the reservoir. Boaters were thrilled, the fish throve, and if the lake continued to fill, then the salt would remain at most a flavor problem, complaints about the early death of hot-water heaters notwithstanding. But the lake stopped rising; instead, it slowly sank even though the cities were not taking their full allocation of water. The cities held back for two reasons: to keep from paying interest on the Bureau loan and to conserve lake water. Despite the cities' earlier assurances to Congress that they wanted to pay off the project in full, they did not complain about *not* paying more than the principal and operational costs, even if it meant taking less water than they could. However, despite the Authority restricting pumping to less than 70 percent of the Bureau's yearly limit, Lake Meredith failed to grow any larger.[14]

At the time, repairing the storm damage the dam had incurred in March and May 1973 took precedence over salt problems, and the Authority hurried to find a contractor who could replace the riprap carried off and shifted by high winds and a tornado that had struck the dam. Without the rock armor, water could start eating into the earth-fill of the dam if the lake rose or another windstorm hit. The following comparatively wet years were not especially problematic where the salt content of the lake was concerned. As a result, the Bureau and the Authority did not aggressively seek solutions to the salt problem until 1977.[15]

Because of the geology and politics only a few options existed for solving the salt-seep nuisance. One involved lining Ute Lake with something waterproof; as with dewatering the lake, this option never reached the serious-consideration stage. Deflecting the groundwater through diversion wells might have been possible, but the cost and the lack of geologic certainty as to exactly where and how the water ran beneath the lake and dam made such a plan unfeasible as well as very expensive. Tests run in the spring of 1978 showed that drilling wells just upstream (in the rock layers) of the seeps and pumping the water out worked on a small scale. Flow to the seep dropped considerably, and this seemed to be a possible solution to the brine inflow, aside from the problem of where to relocate the salt water. Surface ponds provided one possibility, allowing the water to evaporate and the salt to accumulate. Investigators from CRMWA and the Bureau also considered selling the captured or diverted brine to industry or using injection wells to pump it into an even deeper rock formation, where the brine would not migrate. By the close of the decade, the question of Lake Meredith's salt inflow and increasing salinity had been answered at least in part. However, one more element contributed to the salinity equation, an element that also partly explained the lake's failure to keep growing.[16]

Aside from the new brine springs, groundwater flows into the Canadian declined greatly over the course of the last half of the twentieth century because of irrigation. Recall that competition with irrigation from the Ogallala Aquifer in the South Plains had pushed Lubbock toward working for the Canadian River Project. The number of wells in the South Plains grew rapidly, especially after the drought of the 1950s reconfirmed the economic benefits of irrigation. However, such practices spread only slowly to the area around the Canadian and the northern Panhandle. One reason was the greater amount of ranching compared to farming. The greater depth of water in the northern Panhandle also made pumping more expensive for farmers until the discovery of the Hugoton natural gas field. Relatively inexpensive gas helped ease that problem, leaving only the lack of suitable terrain to slow the spread of irrigation.[17]

Flood-furrow irrigation, one of the oldest types of irrigation, required almost perfectly level ground so that the water would not run off and erode the land instead of soaking into the dry soil. But the field should not be too level because the water needed gravity to flow and excess water had to drain away so that the crops would not rot or the soil turn salty. The rolling uplands of the Breaks and sand hills north of the Canadian did not lend themselves to flood-furrow irrigation, and no crops in the area brought enough money to justify the cost of leveling such land. So matters stood until Frank Zybach, a farmer in Nebraska, found a solution to the terrain problem that also inadvertently affected the flow of the Canadian River.[18]

Zybach invented the center pivot irrigation system. After the pumps brought water to the surface, Zybach's machines ran the water through elevated pipes mounted on rolling supports that looked a bit like miniature power line towers. The pipe attached to the well at the center of the field, and the sprinkler system rotated around the well, leading to the name "center pivot" irrigation system. Instead of flooding the ground, a center pivot "rained" water down onto the crops. The wheels let it travel over uneven terrain, markedly increasing the amount of Great Plains land that could be irrigated. Zybach patented his design in 1952 and then sold it to a manufacturing company two years later. Although farmers were slow to buy the new equipment at first, it soon spread over the Great Plains. Center pivots, when adjusted correctly, reduced the amount of water that had to be applied to the land in order to grow a crop. Center pivots certainly reduced the labor required to irrigate. Instead of tromping out into the muddy field to move pipes, open valves, and to repair damaged furrows by hand, farmers could set a timer and go home. The machines did the rest.[19]

An inadvertent side effect of this increased irrigation proved to be a drier Canadian River. According to interviews done by hydrologist Gunnar Brune, ranchers and property owners in the Canadian River watershed noticed that spring flows grew weaker beginning in the 1920s and 1930s, a development that accelerated in the 1950s. Groundwater decline accounted for much of this drying, although it was not the only cause. In some cases, upstream wells and the creation of lakes over a few springs disrupted flows, but groundwater use seems to have caused the greatest drops in spring output. Pumping associated with oil exploration and production began weakening or drying up some Hutchinson County springs in the 1920s. Other Hutchinson County springs, such as Bugbee Spring, declined almost to nothing between 1948 and 1958. The cluster of seeps and springs near the old site of Plemons, north and slightly east of Borger, flowed in 1943 but had gone dry by 1980. White Deer Spring, also in Hutchinson County and downstream of Sanford Dam, began declining in 1934, with noticeably weaker flows during "irrigation season" as pumping lowered local groundwater levels. To the west in Oldham County, the 1930s also marked the beginning of the end of many seeps and water holes. Ojo Caballo (Horse Spring) changed location, moving downhill from its original position as the water table dropped. By 1938 Agua Pintada, or Rocky Dell Spring, known for its nearby petroglyph of a water serpent, dried to a mere seep. These losses, taken together, had a serious effect on inflow to the Canadian.[20]

The loss of spring flows cost the river a significant amount of water, although exactly how much remains impossible to completely calculate. The spring at Las Tecovas on the Frying Pan Ranch north of Amarillo in Potter County produced 266,416,120 gallons per year in 1881. Pumping in the area for the city of Amarillo and for ranch supplies drew the flow down to 23,311,411.2 gallons per year in 1924. After plummeting to one-fifth of that during the 1930s, the spring recovered somewhat and produced 20,813,760 gallons per year in 1978. Big Spring in Oldham County flowed at 28 liters per second (lps) in 1907, but produced only 0.84 lps in 1977. According to the numbers found in Brune's work, *Springs of Texas*, springs in the five watershed counties of Oldham, Potter, Hartley, Hutchinson, and Moore could have contributed at least 1,265,564,227 gallons, or 3,883.874 acre-feet, of water per year to the Canadian. This number is a best-case hypothetical, because many of the spring flows no longer reached the river and because flow varied greatly between winter (no pumping) and summer (irrigation season). The difference in known flows in Oldham County alone between 1938 and 1980 amounted to 600,132,330 gallons, or 1,841.7 acre-feet, less water per year reaching the Canadian. This difference equaled 47 percent of the five-county spring flow in 1981. In other words, if groundwater had not declined in Oldham County, and if all that water had still reached the river in 1980, there would have been 4,065.6 acre-feet flowing into Lake Meredith. As it was, the theoretical five-county total of 3,883 acre-feet per year accounted for 3.77 percent of the Bureau of Reclamation's yearly pumping allowance, or harvest, from Lake Meredith. This does not seem like much, but in the first six months of 1986 Lamesa needed 1,872 acre-feet, Slaton consumed 1,298 acre-feet, Pampa used 3,447 acre-feet, and Plainview drew 3,041 acre-feet. The declining springs also hinted of another, less obvious problem for the Canadian River.[21]

Declining water tables dried more than the Canadian's springs and tributaries. Recall that the level of the springs, in the case of the Canadian Breaks, marked where the river had eroded through to the Ogallala Aquifer, causing water to seep out of the aquifer and into the streams and river. The riverbed, because it touched the aquifer, gained some water. Even if no water flowed in the sandy channel, a little digging often produced a seep that people and animals could drink from. And the under-river water levels rose fairly quickly when rains came, at best adding to the stream's flow and at minimum reducing its losses. But as the water table dropped between 1920 and 1980 and beyond, the connection between the riverbed and the aquifer severed. The Canadian had almost always gone dry in places during summer or in drought years, as did many Great Plains streams, because evaporation and bed seepage exceeded inflow. Now the disconnection from the aquifer increased and the river lost larger and larger amounts of water into the sands below the channel as well as to evaporation. When rains came, spring flows increased in winter, or excess water spilled out of Ute Dam, more of it sank into the channel sands rather than flowing into the lake. The lower flows left more sediment in the river channel, potentially blocking what water did try to travel downstream. Over time this slow process caused more and more problems for

CRMWA's water managers. The alterations also meant that less fresh-water entered the lake to dilute out the salts left by evaporation, causing Meredith to grow increasingly brackish.[22]

Although humans worried about the salinity of Lake Meredith's water, the piscatorial denizens of the reservoir did not seem to mind it. This was good, because fish-related recreation played a major role in the life (and justification) of the lake. In addition to native species such as various shiners, crayfish, and channel catfish, the lake came to support populations of walleye, sunfish, crappie, large- and smallmouth bass, and flathead catfish. Those who preferred fly-fishing could try the imported rainbow trout found in the stilling basin downstream of Sanford Dam. On their own local anglers added several species of nonnative minnows to the ecosystem by dumping leftover live bait into the lake. The recreational aspect of the reservoir, while welcome, added another layer of bureaucracy to the Canadian's management: CRMWA oversaw water management of the lake and the National Park Service and Texas Parks and Wildlife Department supervised recreation, grazing leases, and archaeological preservation within the Lake Meredith National Recreation Area. Occasionally the two uses came into conflict, as was the case when fisher folk, eager to provide better habitat for the fish, dumped branches into an area not far from the water system pump intakes. Fishing, boating, swimming, camping, and picnicking drew growing crowds to the Canadian Valley as soon as people found enough water to float a boat and drown a worm.[23]

Park visits increased steadily during the decade as area residents took advantage of all that the Breaks and river had to offer. The combination of water, near-constant breezes, and (mostly) warm summer sun encouraged area residents to come and play at the lake. The Bureau of Reclamation's official project history for the 1970–76 period reported that the number of visitors per year ranged from 1,239,640 to 1,826,080. Visitor numbers declined during the national economic downturn in the latter part of the decade, although 1,055,347 people checked into the recreation area during the first half of 1979. Of course, some of these were repeat visitors, especially fisher folk determined to do their part to reduce the bait-worm population and to prevent piscatorial overcrowding of the reservoir. Swimmers dove off rocks and paddled around the official swimming areas, and probably snuck into the water from unapproved locations as well. Picnic areas on both shores of the lake attracted families and community groups to cook out and relax. A small resort area called "Coronado Cove" on the northern bluffs upstream of the dam provided more genteel recreation and served as the center of a small neighborhood.

Campers, including local Boy Scout troops, hiked and spent nights out under the stars (and in the rain on occasion). Hunters also frequented Lake Meredith in season, searching for wild turkey, deer, geese, and ducks with varying degrees of success. People stalked the shores and the reservoir's surface in search of game or just for fun.[24]

CRMWA's board of directors had taken a boat tour of the lake in July 1968, but they were certainly not the first to wet a hull in the lake's waters. As time passed, boat numbers increased over the 1970s, requiring more marina slots and increased safety regulation and patrols. The boats ranged in size from small sailboats to large houseboats that could be rented for a weekend or longer. Powered craft outnumbered sailboats, in part because of the great sailing skill required on the many days when the winds gusted and swirled down the cliffs and over the water, causing strange currents and dangerous conditions for sailing. As a result Panhandle residents added the term "lake wind advisory" to their store of weather talk. A good reason existed for the regulations and cautions, for watchful eyes during children's fishing derbies and in waterfowl hunting season. Just as when the river had flowed free, Lake Meredith claimed lives, something duly reported in the Authority's quarterly meetings and the Bureau of Reclamation's yearly reports.[25]

The Canadian River's record as a dangerous stream continued even after it had been corralled behind Sanford Dam. People overestimated their abilities as swimmers and sailors and underestimated the hazards of deep water. In July 1973 the CRMWA minutes record two deaths: a five-year-old girl drowned at Bates Canyon and a fifty-nine-year-old man had a heart attack. Drownings became a sadly regular happening in the spring and summer. Two more people died in July 1974, and sometime between July and October Sam Roberts's boat ran him over and killed him. Three people drowned in 1975 and one on the July 4th weekend in 1976, "despite the presence of lifeguards." According to the Bureau of Reclamation, a total of twenty-nine people drowned in the lake between 1965 and 1976. Other accidental deaths included heart attacks, and in one case a boy died after digging himself a cave. The cave collapsed, triggering a landslide and killing him. Hunting season brought its own excitement, and one imagines the Authority's board members wincing and shaking their heads at the report on January 14, 1974, of a hunter who had been wounded while he was up in a tree when another hunter mistook him for a wild turkey. Humans were not the only mammals claimed by the river-turned-lake. During the winter of 1972–73, sixteen cows died after falling through the ice on the lake during a winter storm; boaters found the remains several months later. The cattle deaths served as a reminder that for some people around the river, its hazardous nature had not changed since the founding of their ranches decades before.[26]

Ranching continued around, and to an extent within, the Lake Meredith National Recreation Area much as it had before the creation of the lake. Some of the area within the boundaries of the recreation area and Alibates Flint Quarries National Monument remained closed to the general public. Bill Saha of the National Park Service (NPS) asked in 1969 to lease some for that land for grazing. There would be four long-term lease permits offered to ranchers for seven thousand dollars per year and several shorter two-to-three month permits worth one thousand dollars. The NPS representative assured the CRMWA board that they would coordinate with the Department of Agriculture's Soil Conservation Service to determine exactly which areas could be grazed and how many head of livestock would be allowed in order to prevent overgrazing. The first proposed lease would be for 320 animal unit months (AUMs) for three years, to be adjusted as necessary. The board approved the lease. Apparently things went well aside from the sixteen drowned cows in 1973 because the lease question remained out of the minutes until October 1974, when the board discussed the need to move and repair fences to control people and cattle in "camping, fishing and picnic areas" and to control cattle in the Bugbee Creek area. Later in the decade CRMWA signed two more leases for 1,500 AUMs with the Coldwater Cattle Company and a 480-AUM lease with Glen Dean. No cases of hunters mistaking cattle for deer appeared in the official records. Truth be told, the ranchers in and around the Canadian Breaks would probably have been happy if careless hunters had been the worst of their problems.[27]

Financial challenges that beset farmers and ranchers during the 1970s affected the regional economy as well. Prices for beef rose between 1970 and 1973, in part due to inflation, in part because a corn blight cut yields and raised the price of corn for cattle feed, and in part because of the general increase in the cost of agricultural products stimulated by government crop-surplus reduction efforts. Higher beef prices pleased ranchers and meat sellers but not consumers, even those in the High Plains where the regional economy depended on agriculture. As former cowboy, rancher, and historian John R. Erickson put it in his book, *Modern Cowboy*, ranchers knew that prices would eventually decline as people stopped buying beef once the price rose too high, as normally happened with the supply-demand cycle. Instead, in 1973, President Richard Nixon froze beef prices. This may have benefited consumers but it wreaked havoc on ranchers trying to plan for the future. Erickson describes the problems cattlemen faced trying to wait out the freeze, holding cattle until they were "over-fed" and meat quality declined. When the freeze lifted, there were "too many" cattle and prices plunged. Another rancher, T. C. McNeill, recalls trying to sell cattle that he'd bought: "the fair cost of the cows we bought in 1973 was a hundred dollars too high in 1974." The oil leases on ranch land helped a number of the beef raisers who would otherwise have gone broke get through the stormy period, in Erickson's opinion. The price collapse also accelerated ranch consolidation, as big ranches grew larger by purchasing failing smaller outfits. The inflation of the 1970s helped those able to hold on through 1975 because property values rose as many people tried to put money into inflation-resistant assets such as land or metals. At the same time, the growing number of dollars in circulation made dollars cheap compared to land, and paying loans became easier for those who could pay. Still, for many area ranchers along the Canadian River, the first half of the 1970s were a period they would just as soon have done without. The old jokes about "I'm from the federal government and I'm here to help you," grew more bitter.[28]

The federal involvement in agriculture served as another example of the increasing federal oversight and regulations that intertwined with both the wet and paper Canadian River. Several times during the the 1970s, the businessmen and officials who made up the Authority's board of directors glanced over their shoulders and noted that the project would be almost impossible to build in 1975, 1977, or later. Among other con-siderations, John Williams observed that if the Bureau built the dam in 1975, it would cost 250 percent as much as in 1963, in part because of the 5.75 percent interest rather than the original 2.632 percent rate the cit-ies paid to the Bureau. Two years later came another reminder of how fortunate the cities had been. Inspired by both fiscal and environmental concerns, President Jimmy Carter cut or restricted the scope of a number of Bureau and Corps of Engineers projects, especially the construction of new dams, striking a serious blow to the already weakening federal water "conservation" efforts. CRMWA looked into the matter, noting that "President Carter had called for a review of every project underway to see if it could meet certain guidelines. If not, funds would be cut and projects terminated." Mr. Payne from Plainview voted against the larger board and in favor of supporting the president, arguing that some proj-ects needed reviewing and that Carter should indeed review them. Other, more localized federal regulations drew longer-term attention and action from the Authority.[29]

The federal government's interest in and regulation of environmental matters increased steadily over the course of the twentieth century. The new rules, designed to ensure cleaner water, purer air, and better treatment for all citizens of the country, caused some new problems for those trying to manage the Canadian River. The additional tests for water quality cost money, and the small laboratory facility adjacent to the CRMWA headquarters building at Sanford Dam soon proved a bit inadequate for the necessary equipment and materials. The lab expanded, and Ashby Lewis, the "sanitarian" in charge of collecting water samples for testing, also became a general water-quality inspector and an ordinance-enforcement officer, able to ticket those he happened to catch violating fish and game and lake-quality rules. CRMWA, as mentioned earlier, found itself having to monitor concentrated animal feeding operations (CAFOs), or "feedlots" as they are called locally, along with oil-well leakage, pipeline problems, the Dalhart sewage treatment plant, and spraying for insects and weeds far upstream of the dam. In many ways, like the federal regulations that it complied with, CRMWA's purview extended across state lines. A proposed nuclear waste disposal facility in Vermijo Park, in north-central New Mexico, occasioned a letter of interest from CRMWA. At least in this way the Canadian River remained a connection between geographic areas, linking the Sangre de Cristo Mountains with the High Plains. Despite three dams, the wet river continued to bind the region into a cultural and watershed whole.[30]

Sanford Dam did its job regulating the wet Canadian, although not without moments of great interest for those involved with the dam's care and repair. Earthen dams are prone to seepage; so long as the seepage remains clear, all is well. If the water turns brown, engineers get nervous. Seepage is the nature of the structure, and Sanford conformed to the pattern as new springs and seeps appeared almost as soon as water began backing up behind the dam. After close examination the engineers assured everyone that the seepage came from the rising water table behind the structure and not from anything getting through or around the dam. The grout and heavy packing on the northern side and under the dam's foundation held; no brown water appeared to warn of the dissolution of the dam or its foundation. Less "average" were the winds that smote the dam in March and April 1973. The afternoon and night of March 13 storms thundered over the Texas Panhandle with winds gusting up to 80 miles an hour for a period of 36 hours, along with hail, heavy rain, and tornadoes. High waves battered Sanford Dam and stripped away between 700 and 800 cubic yards of the stone riprap covering the upstream side of the dam above the water line. Another storm on April 19 with winds over 100 mph did further damage as well as stealing 750–1,000 acre-feet of water by evaporating it away. The bare area left in the dam's riprap armor by the two storms stretched 2,400 feet (732 m) wide and 35 feet (11 m) up and down. On May 14 at a special meeting the CRMWA board voted to spend $60,000 on repairs, and the work was finished in February of the next year. Aside from these matters, the dam itself caused little concern for the Bureau of Reclamation or CRMWA or the Park Service. The paucity of records in the Bureau's archives concerning the Canadian River Project testifies to the lack of excitement. Considering the unfortunate experiences the Bureau was having with some of its other western water projects, most infamously the collapse of Teton Dam on June 5, 1976, the regional manager probably congratulated himself on having such a quiet, possibly even boring, project. However, while the dam sat firm, the aqueduct kept John Williams and the other employees of CRMWA from worrying about boredom.[31]

The pipeline experienced some problems, including dug holes, blown holes, an interstate, and a lawsuit. The most exciting moment came on November 17, 1977, when a seismograph crew from Shell Oil set off a charge of dynamite a wee bit too close to the main southwest aqueduct. The detonation blew an 18-inch (46 cm) hole in the 27-inch (69 cm) pipe, no doubt inspiring the head of the exploration crew to utter rude words as water poured out of the hole. Shell Oil paid almost $24,000 dollars to replace four 16-foot-long (4.9 m) sections of pipe. A few years earlier, in the summer of 1972, a contractor digging a hole for a new or enlarged cesspit for the Paymaster Gin near O'Donnell had also fractured the pipe. It cost almost two thousand dollars to repair that particular hole. After some discussion and threats, the cotton gin paid much of the cost. Also in the southern part of the pipeline, the extension of the new Interstate 27 caused a year of negotiation and disputation with the Texas Department of Transportation concerning how to protect the aqueduct at Lubbock during and after the construction of the new traffic artery. Since neither the giant pipe nor Lubbock's water treatment facility could be relocated, after much back-and-forth discussion with engineers, workers constructed very large concrete boxes over the pipe. The boxes protected the water lines while still allowing workers to reach the pipes in case of leaks. Meanwhile, erosion caused by heavy rains forced CRMWA and the Bureau to repair and do erosion-control work on a stretch of the line going to Pampa where it crossed White Deer Creek in 1973 and 1975. The river's senior engineer also caused some problems for the Authority's maintenance crews[32]

While the traffic artery represented a very modern phenomenon, the presence of beaver marked the return to an earlier time in the river valley. Hispano settlers in the area had noted and made use of beaver ponds and dams, as did early Anglo residents. The Canadian River Valley lay on the edge of the beaver's usual ranges and would have provided a marginal living for the large rodents. Deforestation during the late 1880s and 1890s finished pushing beavers out of the valley, assuming that the mammals had not already been trapped to the point of local extirpation. So it came as an unwelcome discovery for John Williams and his staff to discover that a family of beavers found the outlet of the stilling basin to be a most congenial habitat. Clearing their dams out of the discharge channel became standard procedure. Deer also returned to the valley, as evidenced by the harvest taken each fall by hunters. Wild turkeys made their home in the brush and under trees, and like the deer ignored fences to venture onto the ranches around the national recreation area. If any prairie chickens were found they did not make the federal wildlife reports, and neither did terrestrial predators such as (ubiquitous) coyotes, bobcats, mountain lions, or black bears. Most people assumed that the latter two remained only in remote areas of New Mexico. Bison also failed to return to the valley, although pronghorn remained on the plains to the north and south.[33]

Smaller mammals also did well, judging by the numbers that masters-degree candidate David P. Hill trapped as part of his summer research in 1978 and 1979. The grasshoppers outnumbered mice and wood rats that year, however, and "[the grasshoppers'] presence on the vegetation was very noticeable," as well as upsetting animal census efforts by tripping traps meant for small mammals. Hill observed coyotes and swift foxes, badgers, ringtail cats, and raccoons (introduced). The beavers that he found upstream of the dam seemed to concentrate in Bonita, Coetas, and Chicken Creeks, favoring Bonita over the others and living on saplings and cattails. The biologist wondered if the large rodents were moving in from the east along with muskrats. Waterfowl too made use of the lake, especially when the shallow playas froze over in winter. The Bureau proudly reported in 1977 that the Fish and Wildlife Service counted: 41,586 mallards, 4,400 widgeon, 630 common mergansers, 505 teal, 250 pintail, one bufflehead duck, and 260 Canada geese on a survey flight in January 1976. An earlier flight, in 1973, produced a less precise total of "65,000 ducks, 15,000 geese and 8 bald eagles in the immediate vicinity of

Lake Meredith." The passage of the seasons and years brought migrating geese and ducks, cranes, and other waterfowl to and past the Canadian Breaks. In much the same, seemingly inexorable way, a human argument slowly reached a settlement.[34]

Not unlike the migratory birds making their rounds through the new lake, the lawsuit over the billing system made its way through the court system. It too would pass through the area multiple times before reaching a landing place of sorts, one not all that far from where the suit started. The courts needed to decide two basic matters: was the 1960 contract binding, and could it be overridden by less than a unanimous vote? Amarillo, Pampa, and Borger filed their suit in 1972, and in December 1973 Fourth District Court Judge Max Boyer ruled that "the original contract of the 11 member cities should stand until unanimously changed by all of the CRMWA members" and thus that all charges should follow the 1960 contract and use proportional billing. Amarillo, Pampa, and Borger were quite pleased with this decision, as one might expect, because Lubbock's 1970 proposal cost Amarillo an additional $20,000 per year, while Borger's bill tripled. Borger had paid but under protest. The judge's decision did not please the southern cities.[35]

After some discussion, CRMWA and the eight individual southern cities challenged the ruling, arguing that the table of percentages shown in the 1960 document applied only to construction of the dam and aqueduct. Other costs were negotiated individually on a pro rata delivery cost. On November 7, 1974, Justices James A. Ellis, Mary Lou Robinson, and Charles L. Reynolds of the Seventh Court of Appeals ruled that each city had to pay the actual cost of maintenance and delivery of water to the city. The appellate court also ruled that the question of unanimity being necessary for accounting changes was a separate case and remanded it back to the Fourth District. Lubbock's newspaper coverage of the decision contained a rather large dollop of pleasure in the results, painting Amarillo and the northern cities as being a touch selfish in their desire to pay less than their fair share. The final court-approved plan, the one J. Ray Dickey of Lubbock had proposed in 1970, reduced Lubbock's costs by $6 million. The O&M costs were to be charged on the basis of water distributed, or in other words CRMWA had to "determine the Operation and Maintenance allocation by actual cost of delivery of water." Which raised the question, once again, of exactly how that was supposed to be done. Amarillo, Borger, and Pampa filed their appeal while Amarillo continued negotiations with Lubbock in order to reach some kind of compromise.[36]

Another lawsuit temporarily drew CRMWA's attention in 1974–75.

This once involved the cities of Brownfield and Lamesa and the Texas Water Quality Board and Texas Water Rights Commission. Brownfield and Lamesa sat within the Colorado River watershed and disposed of their wastewater through irrigation. Or at least, they were trying to. The Water Quality Board ruled that irrigation remained the best waste disposal method and in fact required the cities to use irrigation disposal in order to protect the quality of the Colorado River. However, the Texas Water Rights Commission ruled that the towns could *not* irrigate with the water because that use did not exist in the original permit. Understandably confused, the cities approached CRMWA's entire board for clarification and assistance after the Water Rights Commission ordered the two municipalities to obtain irrigation permits: in other words, to obtain specific irrigation water rights to the Canadian from underneath CRMWA's "umbrella" water rights. Executive director John Williams agreed that the situation would be better for CRMWA as a whole if Lamesa and the others got the irrigation permits. The permits would add a level of protection to CRMWA's rights in case of someone invoking a "higher use" such as irrigation in order to claim water from the river. The Authority was having enough problems guaranteeing their supply as it was.[37]

The Canadian River Compact granted Texas a set amount of water storage but did not limit this to CRMWA. As a result, when the state granted permits to other bodies such as the city of Dumas or the Southwestern Portland Cement Co. of Dumas to build a reservoir, the new permits ate into the amount of water that CRMWA could store behind Sanford Dam. State law declared municipal water rights inferior to irrigation, industry, and domestic use, leaving the Authority in a bind when it tried to defend the member cities' claims to all water in the river and its tributaries. Since the lake was not rising as forecast because of drought and reduced groundwater inflows, every drop became more precious, and the Authority worked hard to protect the interests of the eleven member cities, even as the board members and the cities they represented argued among themselves.[38]

While they awaited a ruling from the state Supreme Court, Amarillo and Lubbock tried to find a compromise. One problem stemmed from elections: every time a new city commission came into power in Amarillo or Lubbock (or Plainview, for that matter), John Williams had to meet with them and explain things all over. At least once a compromise had seemed in hand, but city elections delayed the decision and so the process started anew. Finally, the Texas Supreme Court upheld the appellate court in the spring of 1975 and ordered CRMWA "to charge each city its actual cost of delivering water to them as the cost of operation and maintenance," leaving the construction cost division to the district court to decide. In October a tentative compromise between Lubbock and Amarillo found Amarillo agreeing conditionally, so Lubbock's board members sent the idea back to the city's lawyer for consideration. In 1976 the city of Plainview once more tossed a joker into the deck by deciding that it did not care to send money to CRMWA until it knew what the cost would be, even though Plainview would get money back whichever way the district court ruled. Plainview's board representative said that "if Plainview was assured of receiving the refund, it was embarrassing to him for the city not to pay [the bill]." After more discussion and rewriting part of CRMWA's operations manual, the cities and CRMWA settled their differences in 1977, just in time to fend off a possible new challenge to their rights to the Canadian River.[39]

Was the Canadian part of the Red River watershed? A quick glance at a map determined that the wet river was certainly not, but what about the paper river? Recall that wastewater discharge from Plainview entered into the Red River watershed. In April 1977, the Red River Water Authority (RRWA) sought permission from the state to expand its coverage in a way that included part of the Canadian River watershed. On the surface, the RRWA's request for expansion raised no objections, but John Williams pointed out a legal-precedent problem. "If their boundaries are determined to cover the Canadian River watershed, then some court might later decide that the Canadian River was a tributary of the Red River," with predictably bad results for CRMWA. The board voted its unanimous opposition to the proposal.[40]

A decade of leaks and lawsuits drew to a close in 1980. In some ways the decade ended quietly. The lake rose a little but generally fell, not coming close to the 1973 record of 104 feet and closing out 1979 at 75.99 feet of depth, or 38 percent of "normal conservation capacity." Fishing, boating, hunting, and swimming continued at the recreational area. A mysterious fish kill on 2.5 miles (4 km) of the Canadian 3 miles (4.8 km) upstream of the dam on September 30, 1979, remained mysterious and as of January 9, 1980, no cause had been found but neither had any more fish died. For the men in charge of managing the Canadian River in Texas, the 1970s probably looked like a success despite the arguments and problems. They had a dam and a water supply system that worked. No floods had imperiled Oklahoma, the four lawsuits involving CRMWA or the Bureau had all been resolved, and people of the area made use of the lake and adjacent lands as had been intended two decades and more before. If the Canadian

River was no longer the river it had once been, free-flowing with a valley dotted with trees, plum bushes, and wild grapes, the trees were coming back with help from the Park Service's reforestation and fire-suppression work, deer and beavers could be found once again, and the river made a positive contribution to society. Or at least, a positive contribution as the people in charge of managing the river viewed it.

But the Canadian refused to remain static. In the 1940s, the Bureau and Army Corps of Engineers had expressed doubts that New Mexico would build Ute Dam. New Mexico finished the structure in 1966. Over the next ten years, that structure would become a bone of contention between New Mexico and Texas as water-hungry communities in eastern New Mexico planned to tap the river and the state raised the dam in response. The nonhuman environment would also force CRMWA and area residents to reconsider the role of the Canadian River and its waters.

Chapter 9

PAPER PLUS WATER MAKES (LEGAL) FIRE

n the summer of 1845 Lt. James Abert (U.S. Army) set out on a mission: to determine the exact course of the Rio Colorado, better known as the Canadian River. One hundred forty years later the U.S. Supreme Court pondered a similar question: where did the waters of the Canadian originate? According to the New Mexicans, the water contracted to Texas in the Canadian River Compact all came from downstream of Conchas Lake, excluding any water that spilled out from Conchas Dam or the Arch Hurley Irrigation District. Neither did those spills count toward the maximum 200,000 acre-feet that New Mexico could hold back at Ute Dam, since the water originated upstream of Conchas. Texas, on the other hand, wanted those overflow waters counted because the newly raised floodway gates at Ute Lake trapped the waters within. Once again, arguments raged over definitions and water.

Between 1980 and 1991, the Canadian River continued changing as dams, decreasing inflows, and increasing water usage affected the wet river. At the same time, two legal cases took shape that would determine the volume and shape of both the wet and paper rivers. Outside of the Canadian River watershed, national debates about preservation, conservation, wise use, and exactly which "public" owned the country's "public lands" ebbed and flowed. In many ways, the 1980s served as a prelude for the events of the next decade. Over the decade a wave of litigation and debate built up over the increasingly waterless river, rising until the waves broke in the 1990s.

Lake Meredith's depth declined, rose slightly, and then declined again. In January 1980 John Williams, manager of the Canadian River Municipal Water Authority, reported that the lake stood at 75.99 feet (23 m) deep, holding 38 percent of "normal conservation capacity," or what the lake could safely hold. Heavy runoff from snowmelt in the area kept the water level steady in April, but by July evaporation, seepage, and water use by CRMWA's eleven member cities lowered the water level down to 74.78 feet (22.9 m), the lowest since 1967. Water flowed into the lake, but even more water flowed out through the pumps behind the dam or evaporated. Come October 1980 levels dropped to 70.30 feet (21.4 m), 28.5 percent of normal conservation capacity. The trend continued until the fall of 1981, when surges of water from storms in late August and early September bumped the level up 20 feet to a volume of 164,179 acre-feet. The upward surge continued as wet weather both increased lake inflow and reduced water demand by the cities—no one needed to water their lawns. By January 1982, 321,200 a/f of water rested behind the dam, with 243,000 a/f of that available for CRMWA's use.[1]

This happy trend continued through April 1983. Williams noted that the inflow of August and September 1981 had been "the largest received in a comparable period since July 1965." In the past, the Canadian's secondary flood season came in late summer, with the big floods of 1893, 1904, 1941, and other high-water events coinciding with the southwestern monsoon season. A series of El Niño years also helped in the early 1980s. The record snowpack of 1982–83 strained dams all over the West and Southwest, forcing record releases from Glen Canyon and Hoover dams, as well as from Ute Dam upstream of Lake Meredith. In January 1983 Lake Meredith rose back up to 95.72 feet (29.2 m) deep, holding 370,000 a/f of usable water and delighting boaters and fishermen (and bait sellers, the local beer-barn, boat dealers, and other secondary economic beneficiaries). As a result of abnormally high snowfall and releases from Ute Lake, Lake Meredith topped out at 96.94 feet (29.6 m) deep in April. The wet and stormy weather made necessary repair and maintenance work more challenging but no one begrudged the extra effort as long as the lake kept rising.[2]

Then it stopped. Slowly, as the weather patterns shifted and water use increased, the lake began declining. Aside from occasional bumps due to storms and the spills from Conchas and Ute on May 17, 1987, the lake level slowly sank for the rest of the decade. Water quality declined with water quantity as salt accumulated in the lake. In late 1981, when a short drought came to an end, the salt level had been 400 milligrams/liter of water, a level the lake reached again in July 1990. World Health Organization maxima for water recommended no more than 500 mg/l of salt in drinking water. For comparison, seawater is 35,000 mg/l and water with over 1,500 mg/l tastes unpleasantly salty. Amarillo's residents, as we have seen, had not waited for salt levels to reach that level before complaining. As the decade progressed, more and more water users wanted to know where their water was going and why the lake declined. [3]

Howard Eugene Robbins had provided the larger answer to the question in 1949. Recall his letter to all the major water seekers, arguing that the Canadian River lacked the flow necessary to provide the annual 100,000 a/f promised by the Bureau of Reclamation. As happened with the Colorado River to the west, engineers had made the flow measurements for the Canadian during an unusually wet period. Once that wet period ended, the aspirations of the dam builders collided with the environmental limits of the Canadian River watershed and crashed. The water the engineers wanted did not exist. But it would take a while for the area's water planners to fully recognize this fact. Something would have to change: either the water source, or how people used the water. But first John Williams and the CRMWA board of directors opted to look for more localized causes for the lack of water. After 1984, they gained a new culprit to blame in addition to drought and evaporation.

Of all the causes of Lake Meredith's failure to fill, drought and evaporation proved the least amenable to manmade cures. The 1970s started with another near-record low of 9.56 inches (242 mm) of rain in Amarillo. Rainfall rebounded, but 1972–73 and 1976 remained below the thirty-year average, with a decadal average of 18.80 inches (478 mm) at the Amarillo rain gauge. The 1980s averaged a better 23.56 inches (598 mm), even counting dry years such as 1983 (14.98 in or 380 mm) and 1980 (13.39 in or 340 mm). Strong El Niños brought storms and moisture in late 1983 and in 1984, followed by a weak Niño in 1988–89. The number of winter precipitation events peaked in the late 1980s, augmented by a strong La Niña that lasted from August 1986 to February 1988. Translated into daily life in the Panhandle and South Plains, wetter, stormier winters and springs dominated the early part of the decade (rising lake), while drier years and calmer weather occupied the latter years. Drier weather brought greater evaporation rates with it as well as cutting into the river's inflows. For example, in 1989, 56,000 a/f evaporated, 110,000 a/f flowed into the lake, and 68,862 a/f were pumped out. Short of putting a cover over the lake, nothing could be done about evaporation losses. But perhaps the Authority could find a remedy for the new constriction upstream. [4]

Ute Dam had grown. Or more correctly, the State of New Mexico installed floodway gates in the spillway in 1984, raising the height of the dam from 3,760 feet above sea level to 3,787 feet (1,147 m to 1,155 m). This allowed the dam to hold back more water, increasing Ute Lake's depth by 27 feet and raising the volume from 108,000 a/f to potentially 272,800 a/f, if New Mexico opted to hold back all water possible. This was considerably more than the 200,000 a/f "aggregate conservation storage" limit found in the Canadian River Compact. Ute Lake deepened as soon as the gates closed because heavy runoff from a flood remained behind the dam. As per the compact, New Mexico could hold such floodwaters for up to eighteen months. When the waters remained impounded a year and a half later, Texas and Oklahoma requested that the floodwaters be released and the floodway gates opened or (preferably) removed. New Mexico's state government considered the request and refused.[5]

CRMWA's challenge to New Mexico's elevation of Ute Dam centered on the term "conservation storage." The Texas Water Development Board (TWDB) reservoir terminology page defines a dam's conservation storage capacity as "the space available in a reservoir to store water for subsequent release or withdrawal to serve the needs for man's various beneficial uses that is between the lowest outlet level of the reservoir and the conservation pool elevation." In other words "conservation storage capacity" refers to the space between the bottom of the spillway and the reservoir's maximum nonflood elevation. TWDB defines conservation storage as "the volume of water present within the conservation storage capacity, not including any water above the top of conservation pool (cannot exceed the capacity) or in the dead pool storage." Water in the dead pool is water that cannot flow out of the lake without help; it must be pumped out. Another term that appeared over and over in the lawsuit would be siltation pool or "sediment reserve," which meant the volume of the reservoir that had been set aside from the very beginning for the dirt, sand, and silt washing into the lake. The Canadian River Compact did not include these definitions, thus opening up room for legal battles.[6]

The March 6, 1984, meeting of the Canadian River Commission produced a stalemate all too familiar to some members of the CRMWA board of directors. John Williams's quarterly report of April 10 explained, "The Commission meeting did not produce any agreement among the states concerning the meaning of conservation storage, but the New Mexico representative did indicate that the newly-enlarged Ute Lake would not be maintained completely full under current conditions." New Mexico's commission member, Philip B. Mutz, later recounted that the additional storage capacity was for silt control, not conservation storage, and the "minimum pool maintained at 50,000 acre-feet did not constitute conservation storage because it was held for sediment control and recreation and was not available for release for the purposes specified in the compact definition of 'conservation storage.'" Since the silt would accumulate over time, filling the extra capacity, New Mexico argued that the downstream states should not be concerned by the change. Indeed, earlier studies reported that upstream of Ute Dam, Conchas had already lost roughly 2,030 a/f of capacity per year between 1939 and 1963, so silt capacity was a definite concern. The special water master appointed by the U.S. Supreme Court later noted that perhaps Texas and Oklahoma should have raised the question about capacities and definitions even earlier, since the question of whether "water stored or capacity allocated solely for recreational use of water stored in unused sediment control capacity constituted 'conservation storage' under the compact, surfaced as early as 1953 and 1955, respectively, but had not been resolved." Indeed, in July 1984 John Williams stated that he had met with the Texas representative of the Canadian River Commission to sort out how to define conservation storage. "The definition used in the Canadian River Compact seems incomplete since it does not address certain uses of water such as recreation." In short, was the extra space "conservation storage" for holding water (conserving it) for future use, or was it just sediment and recreational water that did not count toward the 200,000 a/f limit?[7]

Because of the need for unanimity, the Canadian River Commission failed to reach an agreement, pushing Oklahoma and Texas to seek a legal decision. After almost three years of discussion, Texas and Oklahoma presented New Mexico's representative with "a motion for a complaint and a brief filed in the U.S. Supreme Court . . . [that] alleged that New Mexico was in violation of the Canadian River Compact by constructing conservation storage in excess of 200,000 a/f." An additional complaint emerged later and centered on the water spilling from Ute in 1987. New Mexico, in a report subsequent to the April–July 1987 water releases, "contended that water spilled or released from Conchas Reservoir into the Canadian River" was not water "originating in the drainage basin of the Canadian River below Conchas Dam" and so did not count toward the 200,000 a/f total. Texas and Oklahoma disagreed with this analysis as well and added it to their complaint to the Supreme Court.[8]

As *Texas and Oklahoma v. New Mexico* inched closer to the top of the Supreme Court's docket, CRMWA watched a second lawsuit developing closer to home. This second case also stemmed from a question of definitions, notably: what and where are the banks of a waterless river? Although sounding at best like a question from a Zen master along the lines of "what is the sound of one hand clapping?" and at worst like a

Figure 9: State oil wells (1926?). Used with permission of the Hutchinson County Museum, Borger, Tex.

way for state-paid surveyors to waste taxpayer monies, the answer held the potential to bring tens to hundreds of thousands of dollars to either the State of Texas or to landowners along the Canadian downstream of Sanford Dam. The decision also greatly affected area hunters and recreational vehicle users. In some ways, the matter looked back to 1926, when the state had tried to claim oil royalties from wells located in normally dry stream beds and washes in and around the Borger Oil Field. According to state law, the bed of a navigable river belonged to the state, along with the water flowing above it and the minerals below the bed. For example, the wells in Figure 9 clearly belonged to the State of Texas, which collected all royalties and fees from the oil producer. In contrast, the proceeds from the wells in Figure 10 went to the landowner, whoever he or she was.[9]

After the closing of Sanford Dam's gates, the river channel downstream shrank just as it had below Conchas and Ute. One reason that the channel shrank stemmed from the lack of floods in the river valley. The geomorphologists Stanley Shumm and Robert Lichty first described this pattern on the Cimarron River in far southwestern Kansas, where a major flood widened the river's channel tenfold in the early twentieth century. High waters scoured away small (and large) trees and shrubs, loosening the sandy soil and reworking sandbars within the channel. The water coming from Lake Meredith's spillway lacked the "oomph" of the river's earlier floodwaters. The same weakness applied to seeps and smaller releases from the upstream dams as well. Without the floods, more plants grew along the river channel. Those plants also grew larger, stabilizing the banks and eventually encroaching farther into the old active channel. Even if the dams released large flows that could "clean out" the vegetation, the water immediately downstream of the dam would dig down rather than spreading out as it had once before.[10]

The reason for the down-dam digging related to how streams move sediment. Riverbed erosion and deposition depended on four variables: the slope of the streambed, the velocity of the water, the amount of sediment already in the water, and the size of the sediment to be moved. The Canadian left its sediment load behind at each dam, making the water "hungry" when it emerged from the spillways. "Hungry" water ate into the bed, cutting a deeper, narrower channel until it picked up enough material to regain that part of the balance. As a result, the river channel narrowed and the banks closed in. Father downstream where the amount of sediment increased, the river shifted to carrying the sediment or even depositing it if the speed of the river's flow decreased or the amount of water declined through evaporation (for example). Because of these

Figure 10: Privately owned wells near Borger. Taken by J. Evetts Haley on March 8, 1926. Used with permission of Nita Stewart Haley Memorial Library, Midland, Tex. Image JEH-Y-30.6/.

Figure 11: The state's view of the Canadian. Taken by J. Evetts Haley on March 8, 1926. Used with permission of Nita Stewart Haley Memorial Library, Midland, Tex. Image JEH-Y-30.14.

effects from Sanford Dam and the other upstream changes to the Canadian, a person might walk up to half a mile from the pre-1965 riverbank until she found the 1986 stream.[11]

Because of these alterations, the exact edge of state-owned and privately held property became increasingly unclear. This riparian uncertainty affected landowners downstream of Lake Meredith for two reasons: royalties and trespassers. Recall that oil-well income from wells in the riverbed belonged to the state to use for all residents, while revenue from wells above the riverbank belonged to the private property owner. In a similar way, hunters and users of the increasingly popular off-road vehicles (ORVs) such as four-wheelers and dune buggies could walk in or drive in the riverbed, since it belonged to the state and therefore belonging to all state residents to use within certain legal limits. However, ranchers did not appreciate people straying onto their property without permission. So where was the property line between the state-owned riverbed and the adjacent ranch lands? The state maintained that the original riverbed from before the dam formed the line, while landowners pointed out that since state law defined the edge of the river as the banks, and the river no longer came close to the 1968 borders, then how could the long-abandoned banks be the edge-of-water?[12]

RED WATER, BLACK GOLD

This dispute over the exact edges of the Canadian River's channel and bed reflected a larger debate going on in the United States in the late 1970s through the mid-1980s, although they were not directly part of that debate. The environmental legislation of the 1960s and 1970s and the environmental movement as a whole struck some observers as having exceeded the limits of federal authority to regulate what an individual or state could do with her, his, or its property, or with property leased from the federal government. While supporting conservation, clean water and air, and other health and safety aspects of state and federal legislation, some private property owners and industrial groups began expressing concern that the federal government had overreached. The Endangered Species Act (ESA) generated a great deal of controversy because in order to protect species of plants and animals from extinction, the legislation allowed federal officials to restrict what an owner could do with her property once that area had been designated as necessary habitat for the endangered species. Additions to the ESA also specified penalties for disturbing that habitat or harming members of the protected species, even unknowingly. Uncertainty about the exact state of land in areas where an endangered plant or animal might be found and whether resources on that land could be used depressed land prices and discouraged third parties, such as logging companies or grazing lessees, from bidding on contracts even for privately held land.[13]

At the same time, especially in the western United States, a general resurgence of social and fiscal conservatism provided a supportive atmosphere for opponents of further (or even of existing) federal efforts to regulate environmental quality and to tighten up land use policies, notably grazing leases and permits. A number of Americans developed a sense of "wait and see" toward environmental protection: the necessary legislation had been passed, air and water quality were improving, so the time had come to stop and see how the trend progressed before demanding more action. For some conservationists the late 1970s seemed to be a time to rest and regroup. The environmental protection and preservation battles of the previous three decades had taken a great deal of sustained effort, beginning with the fight to protect Dinosaur National Monument, then defending Echo Canyon, working for the Clean Air and Clean Water Acts, lobbying for the passage of the Wilderness Act (1964) and Endangered Species Act (1973) and other equally intense local efforts. Some people grew tired of nonstop activism; after all, everything that needed to be done might well have been done. They had saved the Grand Canyon, new laws protected the bald eagle, peregrine falcon, and other species, and Congress had banned DDT. National economic difficulties also reduced the interest in spending for environmental causes, and partly as a result membership in groups such as the Sierra Club declined. Even as the energy of the environmental movement seemed to ebb, the recession and stagflation of the late 1970s encouraged people who wanted to trim (or chop) the costs of overseeing federal projects and spending, including enforcement of a large number of regulations.[14]

The confluence of rising demands for local control and falling state revenues led to the so-called Sagebrush Revolution in Nevada, Idaho, Utah, and other western states with a large proportion of federal lands. Governors and legislators, as well as some citizens of the states, demanded that federal agents of the Forest Service and Bureau of Land Management take more heed of local needs and desires. Failing that, at least part of those federal lands should be given to the individual states or offered for sale to the grazing or mining leaseholders. The income from royalties, leases, rents, and taxes would buoy the coffers of revenue-starved states and counties as well as "getting the feds out." To arguments that federal lands formed a commons belonging to all Americans and that it was necessary to protect public access to public lands, the "Sagebrush Rebels" pointed out that more than 70 percent of Nevada belonged to the national taxpayers, including residents of Nevada, and that turning at least some of that land over to the states or to state residents who wanted to buy it would not unduly injure residents of New Jersey or Ohio. Westerners' arguments over administration and sequestration of federally owned land, such as designating forests or portions of parks as wilderness areas and roadless areas, grew heated during the early part of the 1980s.[15]

President Ronald Reagan's secretary of the interior, James G. Watt, supported and encouraged reducing federal involvement with environmental management, cutting funds to various parts of the regulatory and management sections of the Department of the Interior and accelerating approval of mineral and mining leases. His activities went so far in the direction of deregulation that they helped revitalize groups like the Sierra Club, World Wildlife Fund, and other organizations that had seen their memberships decline or remain steady during the late 1970s. Environmentalists such as David Brower raised the alarm yet again as they watched their hard-won gains seemingly eroded by Watt and the Sagebrush revolutionaries.[16]

Very little federal land existed in Texas, but a similar spirit moved through the air. Texas, because it entered the United States via annexation rather than purchase or conquest, retained all of its state lands. The state government passed homesteading laws, assigned water rights to surface waters, sold or leased some land to the federal government for military posts and national parks, and provided the target for Canadian Valley landowners. When the state attempted to reassert its rights to mineral royalties from oil wells in the Canadian River's pre-dam bed, the ranchers balked. The public arguments of the landowners' attorney, Jody Sheets, aside, the possibility of gaining royalty income probably had more to do with the ranchers' desire to protect their lands from trespass by state tax agents (and hunters and ORV users) than with a desire to protect the rights of landowners from government overreach. Oil prices had begun rising again in the late 1970s and into the early 1980s, inspiring renewed interest in the Panhandle's oil fields and gas wells and the revenue they produced. The Canadian River Municipal Water Authority watched all these matters carefully before venturing to offer an opinion or file a brief.[17]

The Authority discussed becoming involved in the case in October 1989 and revisited the matter in January and April 1990, although the suit had begun almost a decade earlier. In January 1981, John Williams noted in his quarterly report that Huber Oil Co. would drill a well in the floodplain downstream of dam, "apparently in an effort to precipitate legal review of the ownership of that area." As events would prove, no one had ever surveyed the Canadian River's riverbed boundary, leaving the legal "edge of the river" vague at best. Eight years after Huber Oil's notice, the Authority's board of directors noted that it might consider intervening in the pending case and applied with the state district court in Roberts County to determine if CRMWA could participate in the case. The April 11, 1989, meeting of CRMWA's board focused on two pieces of news. First, that the U.S. Supreme Court had appointed a special river master, a neutral outside expert, to try to find a compromise settlement between the three states involved in the Ute Dam debate. The second centered on the downstream suit. "In other litigation matters," the CRMWA records noted,

> the Authority has been found a proper party in the suit regarding land ownership downstream of Sanford Dam; but an Association of local recreational users was not allowed to intervene in this case. The recreational Association is expected to appeal and to file other litigation in this matter. Proposed legislation to allow the sale of land involved and the purchase of other land has been introduced in the Special Legislative session in Austin.

The recreational users, hunters from Hutchinson County, wanted better access in order to hunt on more of the river bottom. They supported the surveys that showed the river to be up to two-thirds of a mile wide, but the court determined that they "had no justiciable controversy" in Roberts County, as Jody Sheets put it. In other words, unlike CRMWA, the hunters had no more interest in the matter than did any other state resident, although those being shut out of their preferred deer-hunting spots no doubt disagreed. Then, fourteen months after the Authority first considered involving itself in the matter, the state courts ordered interested parties to submit maps of where they thought the property lines should be located.[18]

As the question of the current location of the Canadian's banks wound its way through the docket in Roberts County, more current matters demanded the Authority's attention. The eastbound pipeline serving Borger and Pampa sprang a major leak in early summer of 1984. The cause proved to be corrosion, a normal process that had been accelerated by "very salty groundwater. . . . The salt water, likely originating from *oil field* activity in the area, created very favorable conditions for the pipe to act like an anode in bleeding off electrical current to the surrounding soil." Oil producers used brine injection to improve the flow of oil from some wells in a mature field, which the Borger Field had become, by pumping fluids (liquid or gasses) back into the ground as a way to increase pressure in the well to help keep oil flowing. In this case the fluid also inadvertently served as an electrolyte. Oxidation and decomposition occurred at an anode, and so the pipe "rotted" as the electrical activity tore away atoms of the metal, just like an alkaline battery going flat. Workers replaced the pipe and noted the problem for future reference. Indeed, the same thing happened four years later in two more spots along the same pipeline, again caused by brine-accelerated electrolysis. The Authority did not blame any particular oil production company for the developments because just about every producer injected brine and it would have been impossible to isolate the well from which the electrolyte had seeped. Other unusual matters besides water pipes acting like batteries came up in the management of Lake Meredith, adding a bit of color to the board's otherwise routine quarterly discussions of legal matters, water quality, and lake levels.[19]

In addition to the daily round of pumping, scheduled repairs, and administrative details, a few unusual matters drew the attention of the Authority's board members. A major ice storm in late February 1984 knocked out power in Lubbock and disrupted pumping of water to everyone below Lubbock on the aqueduct for a day or two. Three years later the Bureau of Reclamation initiated studies concerning what would happen to Sanford Dam should an earthquake strike the Canadian Valley. The next report noted that the Bureau remained concerned about the same three things its engineers had been concerned about from the construction of the dam: the behavior of observation wells, seepage under and around the dam, and "the old landslide near the river outlet works tower" on the north side of the dam. A geology report in July 1990 noted the possibility for soil liquefaction on the north abutment of the dam in the event of a strong enough earthquake. And in August 1991, "one apparent pipe-bomb was discovered near a structure at Sanford Dam . . . its construction and placement seem to indicate that it might have been an amateurish attempt to destroy fish or perhaps experiment with such devices."[20]

Fires in and around the river added more excitement for the river's managers. Grass fires had swept over the grasslands around the Canadian Breaks long before Anglo Americans settled the area. Because it was moister and had less fire-prone vegetation, the river valley had served as a natural firebreak, as ranchers fighting the great wildfire of 1895 attested. That burn probably began near the Arkansas River in Kansas and burned south as far as the Canadian River Valley, until a snowstorm and wind shift along with the river's waters stopped it. Grassfires and brush blazes continued after the settlement of the valley and the creation of Lake Meredith. Obviously, a brush fire posed no threat to the dam itself, but they occasionally endangered people using the recreational area or visiting Alibates Flint Quarries National Monument and threatened the houses that had begun encroaching on the park. Another concern stemmed from the behavior of fire in broken, steep brushy terrain such as surrounded Lake Meredith. The strong, swirling winds that made sail boating exciting and dangerous at times also caused fires to behave in unpredictable ways, accelerating their progress and at times carrying burning debris over fire lines. Ranchers, Forest Service firefighters, and local volunteers learned quickly the need to stop conflagrations before they reached broken ground. Fire prevention efforts included precautionary burns, and where that was not considered safe, brush clearing by hand. Lightning strikes and human carelessness caused most fires. In late June 1990 the Canadian Valley experienced multiple brush fires that consumed "several thousand acres." Although noteworthy, the burns did not affect water quality enough to merit mention in the various CRMWA and Bureau of Reclamation reports. However, several other water-quality concerns did have to be dealt with during the 1980s.[21]

While the wet river flowed or trickled along its bed and into Lake Meredith, the tap-water river declined in quality and quantity along with the lake level. The salinity of the lake's water rose slowly over the decade after dipping in the 1983–84 wet period. The member cities' residents, especially those whose towns did not blend well water with lake water, expressed concerns about the salt, as did those who had to maintain the equipment through which the brackish tap water flowed. CRMWA, as a water authority, wanted to deal with the problem, but the 1959 enabling legislation did not permit the Authority to take on "salinity control." Therefore CRMWA went before the state legislature in 1987 and petitioned for an addition to the enabling act that would allow the Authority to do what was necessary—conduct research and tests, coordinate with New Mexico and the Bureau on different salt-control techniques—to improve water quality. On May 28 the legislation passed amending the enabling act to better reflect the needs of the water users.[22]

Despite the surge in the lake level, it became obvious that the river could not provide enough water to meet the 103,000 acre-feet per year promised by the Bureau of Reclamation. John Williams and the board and member cities began considering ways to augment the Canadian's decreasing flow. Amarillo already added groundwater from its private well fields for flavor improvement. Lubbock preferred Lake Meredith water "neat" in order to ensure continued good relations with its irrigating neighbors. Instead, the city purchased land and water rights to construct a surface reservoir on the Double Mountain Fork of the Brazos River. South of Lubbock, Lamesa, Tahoka, Brownfield, and O'Donnell possessed neither the populations nor the groundwater to be able to supplement what CRMWA could provide. Increasing flow into the lake by increasing rainfall via cloud seeding never received serious consideration, at least according to the Authority's available records. Weather modification depended on clouds, and when drought hit the Panhandle, it came in the form of brassy skies and crop-withering southwest winds that stripped the air of any moisture, leaving no clouds to seed. Another possible option existed, but it sounded so far-fetched that although the Authority retained some interest in the idea, CRMWA never seriously considered investing in it.[23]

Water, Inc. of Lubbock wanted to import water from Canada or from the Mississippi River to use to recharge the Ogallala Aquifer. The organization, founded in 1967, spent two decades studying the possibilities of recharging the groundwater of the Llano Estacado. Other groups besides Water, Inc., investigated this: the groundwater district near Lubbock, High Plains Underground Water Conservation District #1, sponsored experiments in recycling irrigation water via playa lakes or special recharge wells, with limited success. That was not ambitious enough for Water Inc. Their final draft plan focused on tapping part of the flow of the Mississippi River near the mouth of the river. From there a series of pumps would lift the water across Texas. Some water would go to supply the Dallas–Fort Worth metropolis. The rest would continue uphill and onto the plains, where it would be used to replenish the Ogallala. As shown in Water Inc.'s papers, ten nuclear reactors provided the electricity to run the pumps. It was an amazing idea, although not the most ambitious water-redistribution project discussed at that time. Some entities in California entertained the idea of tapping the Great Slave Lake or Columbia River, channeling the water south over and around the Rocky Mountains and using it to meet California's needs. Back on the east side of the mountains, the states of Mississippi and Louisiana failed to look favorably on Water Inc.'s proposal. The de facto moratorium on constructing nuclear power plants after the accident at Three Mile Island, plus the financial impossibility of growing crops that would bring in enough money to pay for the project, combined with protests from the Army Corps of Engineers and environmentalists, put Water Inc.'s plans on permanent hold. CRMWA's board would have to look somewhere other than the mighty Mississippi for more water.[24]

The end of the decade found the Canadian River in a state of transition, as was its watershed. The enlargement of Ute Dam restricted some flow to the Texas reaches of the stream and stimulated an argument over the meaning of "conservation storage" that sent lawyers for Texas and New Mexico to the U.S. Supreme Court yet again, this time joined by Oklahoma. The spread of salt cedar, the continued flow of brine springs downstream of Ute Dam, and declining groundwater in the Canadian Valley and larger watershed contributed to the declining flows and flavor of the river. Those who depended on the river's waters for drinking and sanitation pushed their representatives to do something to preserve and increase the available water supply, while ranchers and landowners pushed for a definition of the edges of a much-shrunken river. Outside of the valley and the Llano Estacado, Americans debated the limits of environmental protection and federal authority and struggled to balance the needs and desires of a producing and consuming society with the desire to preserve something for plants and animals. The future seemed less certain than it had twenty years before, when a filling lake and splashing fish signaled Anglo Texans' mastery of the Canadian River.

Chapter 10

Two Suits and an Act
of Congress

n the 1990s state and federal courts considered several strange-sound-
ing questions about the Canadian River. Could someone divide lake
water like a layer cake, with 50,000 acre-feet for recreation, 50,000
acre-feet for silt control, 150,000 acre-feet for conservation storage, and
25,000 acre-feet for flood control or fish habitat? If so, then did the
capacity restrictions on a "paper" lake mean anything for the "wet" lake?
And did the dam-induced changes in a wet river affect the banks of the
paper river? The State of Texas answered two of the three questions with
a resounding "no," but the courts did not always agree with the attorney
general of the Lone Star State. At the same time, the managers of the
Canadian River that flowed through pipes from Lake Meredith south to
Lamesa sought more water for their river, reasserting local control over
the federal project in the process.

Local needs and interests drove development and management of the
Canadian River in Texas during the last decade of the twentieth century.
As a result, the Canadian River provided a comfortable living for law-
yers and surveyors as the U.S. Supreme Court and the Texas Supreme
Court both heard cases involving rights to the river or its banks and the
CRMWA began the work necessary to buy itself out of the Bureau of
Reclamation. The major theme of the decade seemed to be one of bot-
tom-up action and agitation for clear definitions of "conservation storage"
and "riverbank."

The first case decided during the 1990s centered on the difference
between water in "conservation storage" and other water in Ute Reservoir
in New Mexico. After Oklahoma, Texas, and New Mexico failed to reach
an agreement over water distribution and storage from Ute Lake in the
mid-1980s, Oklahoma and Texas petitioned the U.S. Supreme Court for

redress. The Court appointed Jerome C. Mays as special master for the Canadian River and charged him with investigating the complaints of the three states. A standard procedure in interstate water cases, the outside, neutral party attempted to sort through the disagreement in hopes of finding a solution that would not require the parties to go back to court. Mays investigated the various aspects of the dispute, met with representatives from all three states, and, after a delay caused by the final negotiations over the Pecos River case of *Texas v. New Mexico*, offered his decision. [1]

Mays reported his findings to the U.S. Supreme Court in October of 1990, and his results did not favor New Mexico. The state could not, in his opinion, separate different portions of the water held behind Ute Dam and declare in effect that "this part counted as conservation storage while this other water did not." Conservation storage referred to the water behind the dam, not the total capacity of the reservoir. Therefore the deepest or most distant part of the lake might be considered additional silt-storage capacity on paper, but water in those parts of the lake depths still counted toward the conservation storage limitations outlined in article IV(b) of the Canadian River Compact. "Recreation" water was not separate from other water either, despite New Mexico's claim. And furthermore, water released from Conchas or returning to the Canadian from the Arch Hurley Conservation District should count toward the 200,000 acre-feet storage limit. However, Mays granted New Mexico whatever waters accumulated in "the dead storage portion of the Ute Reservoir sediment storage pool," that part of the reservoir in which silt collected and that would over time become "dead" as far as the ability to store water in it. Texas and Oklahoma had already agreed to this because almost all reservoir volume calculations ignored dead storage.[2]

However, Mays failed to reach a decision on the desilting pool. According to New Mexico, the water just above the dead storage area constituted a desilting pool where sediment would also accumulate, eventually reducing the amount of water in storage. That silt made the so-called desilting pool in effect a subsection of the dead storage. Thus the desilting pool could not constitute part of the conservation storage for Ute Reservoir. Even if people happened to fish and water ski on the waters there, the desilting pool existed primarily for silt control, and so the capacity (and water in it) failed to apply to the compact's limits. New Mexico also pointed out that part of Lake Meredith operated in the same way, weakening Texas's argument. Texas replied that the two comprised very different cases. In the end, after much study and consideration, Mays sent that particular question back to the Canadian River Commission. Because the states again failed to reach a unanimous decision, the U.S. Supreme Court heard the case on April 16, 1991.[3]

Marian Matthews, New Mexico's deputy attorney general, argued the state's case before the Court. The 200,000 acre-foot conservation-storage limit did not apply to the desilting volume in Ute Lake because the compact's restriction applied to available reservoir capacity, "capacity" meaning specifically that water which could flow downstream without pumping. Water in the desilting pool would not flow out the spillway and so it did not count. Deputy Attorney General Matthews also disagreed with the water master's findings pertaining to waters originating above Conchas and spilling or being released from that dam. As far as New Mexico understood the matter, the word's meaning remained "unambiguous." Since the waters released from Conchas Dam had originally fallen or seeped from upstream of the dam as rainfall, snowmelt, or spring-flow, such waters did not count toward the conservation storage limit below Conchas. However, New Mexico agreed that the question of the exact nature of the "desilting pool" could be settled in the full commission.[4]

Paul Elliott of Texas and R. Thomas Lay from Oklahoma defended their states' claims. While the "so-called" desilting pool might and indeed probably would fill with silt at some point, at present it held water and that water could not be separated from the conservation water described in section IV(b) of the compact. Lay further argued that the special master erred in his finding that Texas and New Mexico's limitations both centered on volume of water in storage. Instead, Oklahoma's attorney insisted that the compact restricted New Mexico's total reservoir *capacity*, not just the volume of water currently in the reservoirs at Ute and Clayton Lakes and another much smaller reservoir. As to originating waters, Oklahoma and Texas both argued that once water left Conchas and returned to the main Canadian River channel, either by going over the floodway or as return-flow from the Arch Hurley Conservation District, that water reoriginated *below* Conchas and so counted toward the 200,000 acre-foot limit. The individual drop of moisture may have entered the stream well above Conchas Lake, but once it passed out of Conchas it became no different from a raindrop falling on the town of Tucumcari or into Ute Creek at Gallegos.[5]

One month later, the U.S. Supreme Court rendered its verdict. The Court began with a brief history of the case and complaints and then addressed Oklahoma's contention that the compact limited available storage and not volume of water. Here the Court found against Oklahoma, based on the language of the compact and the special master's findings. While parts of articles II and IV seemed to support Oklahoma, the weight in article V given to volume and the emphasis in article VIII requiring "accurate records of the *quantities of water stored in reservoirs*," along with "contemporaneous memoranda" from the drawing up of the compact, all pointed toward actual water volume rather than reservoir capacity being the ultimate measure of a lake. Therefore the Court rejected Oklahoma's exception to the special master's report on this part of the decision.[6]

The next portion of the decision split the court, with Justices White, Marshall, Blackmun, Stevens, and Souter ruling against New Mexico's "origination" argument and Justices Rehnquist, O'Connor, Scalia, and Kennedy dissenting. The majority of the Court agreed with Texas and Oklahoma and held that once water left Conchas headed downstream, it no longer originated solely above Conchas, and therefore the water counted toward the 200,000 acre-feet limit. As a result, the spilled water from 1987 should have been let through to Texas and Oklahoma, at least that amount that exceeded the storage limit. The Court pointed out that Texas had a valid point concerning New Mexico's practice of claiming waters that actually originated in Colorado and then flowed into New Mexico, a practice that weakened New Mexico's contention about article IV(a). There was enough ambiguity that the special master, after much study, found that the original drafters, as best he could determine, did not intend for waters flowing out of Conchas to be omitted from the storage limitation. The Supreme Court held that New Mexico's emphasis on one letter from New Mexico state water engineer John Bliss to Senator Clinton Anderson did not support the state's argument, especially in light of additional letters from Bliss to Governor Mabry and other engineers and officials. The speed with which the states drew up and ratified the compact probably increased the air of ambiguity, the Court noted. Also, New Mexico had already conceded that it had all the water it currently needed. "Since the signing of the Compact, there have been no developments in the area below Conchas which require substantial amounts of water for consumptive uses." For all these reasons, the Court ruled that New Mexico's argument was invalid and the upstream state had to include spills and releases from Conchas in the total it could keep back. New Mexico had to release any excess.[7]

The dissenting justices disagreed with this redefinition of "originating." According to Rehnquist, O'Connor, Scalia, and Kennedy, "the Court [majority] concludes that the Compact cannot mean what it says, and instead fashions a different allocation than that which is literally described." Furthermore, "[t]he Court conjures up impractical consequences where none exist," presenting visions of New Mexico chasing down waters that would otherwise flow into Oklahoma and Texas and of Texas being unable to hold water behind Sanford Dam because New Mexico kept so much back. The dissenters held that these visions were simply not true and that the language of the compact was clear: originating meant "coming from above." Even if the water passed through Conchas, it had still begun life (as it were) upstream of that dam and so remained purely New Mexico's to store or distribute as that state saw fit. Otherwise the dissenting justices agreed with the rest of the decision.[8]

The final decision also overruled the special master's finding on the question of the desilting pool. Mays had enough material, the Court found, to have made a decision rather than sending the matter back to the deadlocked commission. Therefore, the Court remanded the case back to the special master "for such further proceedings and recommendations as may be necessary." So in sum, New Mexico lost its argument on origination, Oklahoma lost its argument on reservoir capacity, Jerome Mays had to find a solution to the question of what to do about the capacity in the desilting pool, and Texas got what it wanted.[9]

Interstate water law specialist George W. Sherk pointed out the importance of this ruling nine years later. First, the special master opted to rule in "original intent," trying to go beyond what was written in order to determine what the men who drew up the compact between December 4 and 6, 1950, meant. Most water masters found this unnecessary, given the abundance of definitions and evidence within most river compacts, such as the Colorado River Compact between New Mexico, Colorado, Wyoming, Arizona, Utah, Nevada, and California. Second, as the dissenting justices forcefully pointed out, an interstate compact equaled a contract, and just as in other contracts, the parties had to show that they had "actually been harmed," something Texas and Oklahoma succeeded in doing with their evidence and arguments. In contrast, during the initial 1907 suit between Kansas and Colorado over the Arkansas River, Kansas failed to persuade the U.S. Supreme Court that it had been harmed by the lack of water more than it had benefited from the regional economic boost caused by irrigation upstream. Sherk also used the Canadian River case to further illustrate one of the three ways that interstate water disputes could be resolved: litigation, legislation, or compact negotiation. Although not as well-known as other interstate compacts, the Canadian River Compact and subsequent case marked important legal ground.[10]

What did this mean for the wet river? While the three states and the special master sorted out what to do and what penalties New Mexico owed to whom, above-average rainfall in the early 1990s helped keep Lake Meredith fairly steady. In 1991 131,000 acre-feet of water entered the lake and 120,000 acre-feet departed via pumps (70,000 a/f) and evaporation (50,000 a/f). More than two feet of rain at the dam had helped, as had runoff from other rains and snows in the watershed. By July additional rains bumped the lake up over 6 feet from the start of the year, to 86.65 feet (26.4 m). The lake level declined slightly the next year to 83.49 feet (25.5 m) by January 1993, then dropped to 79.49 feet (24.2 m) that October. Even the first release of water from Ute in September failed to halt the decline. A second release the next spring brought the reservoir up 2 feet (0.61 m). Under orders from the special master New Mexico released a total of 25,000 acre-feet from Ute Dam in penalties, plus whatever exceeded the 200,000 a/f limit. This release helped reduce Lake Meredith's salt content and raised the dropping water levels for a little while longer.[11]

As overdue (at least in some minds) water eased into Lake Meredith, slowing the reservoir's decline and diluting the brine, CRMWA's member cities faced the fact that the Canadian could not deliver what the Bureau had promised. In October 1990, a memorable year that would prove to be the driest in the decade, general manager John Williams reported that since January a total of 58,762 acre-feet of water flowed into the lake as 58,329 a/f went out through the pumps and an additional 48,941 a/f evaporated. Evaporation stole almost as much water as the pumps lifted out of the reservoir. Some years brought more moisture than did others, but individual wet years could not reverse the overall trend. Residents of the eleven cities in the Canadian River's "new" watershed would have to decide what to do about their hydrologic future. They could curtail their water use in order to try to make what the lake held last as long as possible, or they could find other sources of water as a group, or possibly go out on their own as Lubbock and Amarillo tended to do. There was the option not to do anything and instead to hope for a miracle in a region that purportedly received all of three inches of rain during Noah's flood.[12]

What had become of the promised 103,000 acre-feet of water per year? Far less groundwater flowed into the Canadian, and the dropping water table actually pulled some water out of the river when it did run after storms or spring snowmelt. The lack of flow from New Mexico also contributed to the reduction but not in the way that Panhandle residents usually thought about it. Before the dams, flood flows had scoured the channel clear of sandbars and obstructions. Although the "hungry" water released from Ute did cut a channel, it lacked enough force and volume to rework the riverbed sands farther downstream. Partly because of this and partly because of the effects of evaporation and lost groundwater flow, the river wanted for its former power and more water seeped into the bed, moistening the sand but not reaching the lake in its former quantities. Evaporation probably increased during the 1980s and early 1990s as global temperatures increased, although it was not as warm as the 1930s or as dry as the 1950s had been. The effects of hundreds of small dams built by individual ranchers to trap runoff for their cattle to drink contributed as well. For all these reasons and probably a few more, the wet river delivered an average of only 76,000 acre-feet, of which the eleven cities could use 69,000 a/f. The wet, gaining years did not match either the losses in dry periods or the increasing domestic and industrial consumption of water on the Llano. CRMWA opted to find another water source.[13]

When faced with a similar situation, other municipalities chose differently than did CRMWA's membership. Santa Fe, New Mexico, dealt with water shortages in the 1990s due to ongoing regional droughts. Santa Fe, perched on a plateau in the high desert not far from the Rio Grande, found itself backed into what might be termed a hydrologic corner by geography and the law. Beginning in the early 1980s, the formerly small, rather sleepy state capital experienced a surge both in visitors and in permanent population as the city became a center for art and culture. The city drew on groundwater for its needs and soon found the usual sources running nearly dry. Without rain to recharge the shallow aquifer, and with access to the rivers and streams blocked by prior appropriation of the surface water for irrigation and domestic use, the City of Santa Fe began legislating mandatory conservation starting in 1996. The new arrivals would have to live with what long-time residents knew—turfgrass, old-style toilets, private swimming pools, and large showerheads were potential liabilities. Regulations on landscaping, plumbing codes, and wastewater disposal made water restrictions a way of life. Posters in state and municipal offices and cards in hotel rooms, along with articles in the local newspapers, explained to visitors and residents alike the realities of living in the high desert.[14]

Back on the Llano Estacado, residents chose to return to groundwater or extended surface-water holdings. Amarillo and Lubbock had never relinquished their municipal water rights, either underground or on the surface. Since 1969, Amarillo blended some of its groundwater with the Canadian River water CRMWA supplied, and the city simply increased the percentage of well water in the mixture delivered to customers' taps.

Lubbock chose to go a different route and closed the gates of the dam of what would become Lake Alan Henry on the Double Mountain Fork of the Brazos River in 1993. Located near Post, Texas, these impounded waters served as a secondary reserve as well as providing fishing and camping opportunities. Recall that the groundwater in the South Plains had declined more quickly than did the Ogallala near or north of Amarillo, affecting Lubbock's groundwater holdings as well as the farmers' supplies. Farther south, Lamesa and Tahoka lacked both groundwater reserves and the financial resources of the larger cities. Along with the other smaller communities such as Brownfield, Levelland, and O'Donnell, the southernmost CRMWA members worked hard to survive on agriculture- and oil-based economies, augmented by government and some tourism and service industries. CRMWA supplied their only source of water, and if the Canadian failed to provide, something else would have to.[15]

Why not conservation rather than groundwater acquisition? It is possible that the culture and background of the decision makers, including the CRMWA directors and the city commissions and other leadership of the member cities, predisposed them to focus on increasing supply rather than trimming demand. The water history of Anglo Texan residence on the High Plains centered on groundwater and local surface water and the increasingly efficient use of both. Municipal use and discussions in Lubbock and Amarillo during much of the twentieth century highlighted seemingly plentiful supplies and extolled using the water to make the plains bloom and to provide amenities such as shade trees and turf-grass lawns. According to this view, after initial difficulties Amarillo always had plenty of water aside from a few periods when demand temporarily outstripped supply, and the cities always met those needs by buying more groundwater and laying more pipes and by building a dam or two. John Williams worked with river management his entire professional career, first with the Bureau of Reclamation and then, beginning in 1968, with CRMWA. Most other area residents were familiar with the Ogallala Aquifer through their business dealings or the frequent news reports on irrigation and commodity prices. They heard reports that they would run out of groundwater in 1970, in 1980, and so on, but while groundwater levels declined, plenty still seemed to be left, and people came to look dimly on the hydrologists who cried wolf. After the internal disputes of the 1950s and 1970s, CRMWA's leadership might not have felt strong enough to persuade the leaders and other residents of eleven towns to agree first to water restrictions and then on penalties and procedures. Amarillo and

Lubbock tended to go onto voluntary water restriction when necessary, so perhaps at the time that seemed enough to ask.[16]

There had been another option. Recall that Water Inc. of Lubbock proposed importing water from the Mississippi River and pumping it onto the Llano in order to recharge the Ogallala Aquifer. CRMWA might have bought some of that water had the project ever gone through, but the combination of economics and environmental concerns doomed the plan. If the prospect of paying fifteen cents per one thousand gallons of water (wholesale) had made town residents flinch in the 1950s, the idea of paying upward of two dollars or more per thousand gallons would have sent them running to the state utilities commission to protest. As it was, Water Inc.'s dream of a massive water transfer remained just that, a dream, and one that CRMWA had chosen not to participate in.[17]

Instead, the Authority looked east to Roberts County downstream of Lake Meredith. Rough, broken, and scenic, the county had an economy derived from oil production and some tourism, but mostly ranching. In 1990 the county boasted a population of 1,025 people, of whom 675 lived in the county seat of Miami (pronounced "My-amah"). Local historian and author John Erickson noted that as of the 1980s, the county remained so rural that telephone lines still had not reached north of the Canadian River. Because of the broken terrain and emphasis on ranching and oil and gas production, people used very little groundwater. This area received greater rainfall than did the counties to the west. Because everyone ranched rather than farmed, Roberts County residents had a little less need for groundwater, and CRMWA stood less chance of irritating farmers if the Authority purchased some groundwater rights.[18]

And so the Canadian River Municipal Water Authority invested in groundwater. During the 1970s Southwestern Public Service, an electricity production and distribution company headquartered in Amarillo, had purchased water rights in Roberts County to ensure supply for a project that never came to fruition. As a result those rights became available on the market, and in 1994, the Authority's board of directors voted to purchase 42,765 acres of water rights in Roberts and Hutchinson Counties. During the next year CRMWA worked with the Texas legislature to obtain two amendments to its enabling act that would allow the Authority to purchase and pump the groundwater, and to issue revenue bonds in order to pay for the purchase. The local groundwater district, the Panhandle Groundwater Conservation District, approved the sale, and in January 1996 the district issued a permit allowing CRMWA to pump up to 40,000 acre-feet per year from its holdings. Eight months later the Authority sold $19.5 million in revenue bonds, to be paid for through water sales, in order to finance the hydrologic investigation and purchase of the permitted water rights. While this bill seemed daunting, the U.S. Congress helped ease some of the pressure by passing a drought-relief measure in October that gave the Authority a $4 million credit with the Bureau of Reclamation and allowed the Authority to defer payments for three years in light of the ongoing drought and the lack of water available to the Canadian River Project. The financial breathing room permitted CRMWA to begin water deliveries and to start paying off the bonds without having to pay the Bureau at the same time, giving the Authority a small cushion of time and resources.[19]

However, in February 1997 CRMWA discovered a problem with this plan even before the Authority began obtaining funds for the water rights purchases and necessary infrastructure. The Bureau of Reclamation did not support CRMWA's action. Relations between the Authority and the Bureau had been cordial for most of the life of the project, going back to A. A. Meredith in the 1950s. In October 1989 the Bureau's office in Amarillo closed as part of a consolidation process and its functions transferred to Oklahoma City, while the Great Plains regional headquarters remained in Montana. When Williams proposed the groundwater addition, the Bureau decided to shift that part of the project to the Austin office. As a result, the new administrators were less familiar with the overall needs and purpose of the Canadian River Project, and the Bureau balked. Nothing in the contract between the cities and the Bureau allowed groundwater to flow through the system. As John C. Williams explained to historian Rusty Hawkins, "her [the Bureau representative's] reaction was to send [CRMWA] a letter stating that they could not allow us to put groundwater from our new project into the aqueduct they had built for us back in the 1960s because Congress didn't authorize that project to carry groundwater." Or, as the letter put it, the Bureau lacked "adequate authority to allow the use of Canadian River Project facilities for the storage or conveyance of non-project water."[20]

Faced with a nearly immovable object in the form of the Bureau, John Williams and CRMWA's members became an irresistible force. They initiated a two-part process, working with the Bureau of Reclamation while pushing for congressional emendation of the 1950 bill authorizing the project. On June 20, 1997, Representative William "Mac" Thornberry introduced H.R. 2007, a bill that added groundwater provisions to the original act of December 29, 1950. John Williams and the CRMWA board also negotiated with the Bureau for "permission" to add groundwater to the river water in the pipelines. No problem arose from mixing "federal" water with "private" water because CRMWA held the rights to both from the State of Texas. Instead, the Bureau required the Authority to perform additional work along with the engineering review necessary to ensure that the new pumps and pipeline would not cause problems for the main aqueduct. At the Authority's expense, an outside expert conducted an environmental assessment to ensure that the new work would meet the National Environmental Protection Act (NEPA) requirements for a Bureau project. CRMWA also agreed to pay the Bureau roughly $18,200 in administrative costs for a memorandum of understanding (MoU), and the Bureau would perform a "review . . . [of] proposed plans, verification of easements, review of all necessary regulatory permits" and other necessary matters, with CRMWA paying for everything and assuming all liability. And even then, John Williams told the House Water and Power Resources Subcommittee in July 1997, the Bureau was not certain it could authorize the groundwater addition. Thus the need for H.R. 2007.[21]

Williams and CRMWA used two arguments in their quest for groundwater. The first was quantity. But they also used quality. While safe to drink, with 400 mg/l of salt in it Lake Meredith water failed to meet Texas's water-quality standards. Williams made this second point in his testimony, trying to show that CRMWA was doing all it could on its own and working with state and federal agencies and groups in order to provide clean water that met all state and federal minima and maxima.[22]

When legislation failed, the Authority turned to lucre. The bill passed the House but failed in the Senate. As a result the CRMWA board decided that John Williams could take whatever actions necessary in order to get the new water through the pipeline. After considering some other options and ideas, the board decided that if the Bureau would not allow CRMWA to run the water, then CRMWA would buy back the entire project thirty years early. The buyout would not be inexpensive, especially coming so soon after the Authority had floated the large bond issue to buy the water rights, but early purchase seemed to be the only solution. As Williams explained, his efforts benefited from good timing because "the Bureau at that time was very interested in what they called 'title transfer' [because] Congress was wanting them to give up any management rights that they had on local projects" as a way to cut costs. John Williams, Hal Miner of Amarillo, Steven Tucker from Slaton, and CRMWA board president E. R. Moore all flew up to Missoula to see what would be required to purchase the project back. As historian Rusty Hawkins put it, "they found another dead end." The Bureau informed the four men that the Bureau would not take their money because nothing in its contracts or authorizations for the project included early repayments. Since the historic mission of the Bureau, in theory at least, was to help individuals establish themselves on irrigated farms that would be paid off within ten to twenty years, the unwillingness of the Bureau to accept the offer of cash at best seems confusing and at worst serves as an example of bureaucratic entrenchment and fear of lack of precedent. Since no one else paid off their projects early, and no explicit provisions in the 1950 and 1960 documents and contracts allowed CRMWA to finish paying before 2030, then why should CRMWA be allowed to pay? One also wonders if the Authority's practice of consistently delivering less than 70 percent of the yearly water allotment in order to keep from having to pay the full interest due on the project made the Bureau less inclined to accept cash, but that is pure speculation.[23]

In the end, local determination plus political leverage overcame the Bureau's resistance. John Williams turned once again to Representative Mac Thornberry (Amarillo-area), buttressed by representatives Larry Combest (19th District–Lubbock) and Charlie Stenholm (17th District–Abilene) for assistance. In mid-1998 legislators combined the Canadian River Project, originally a separate piece of legislation, with projects in California, Idaho, Arizona, Texas, New Mexico, and other states into H.R. 4389, a bill "to provide for the conveyance of various reclamation project facilities to local water authorities." Under title VI of the bill, the Authority would repay the Bureau $34,806,731, the Army Corps would take over running the flood control aspects of Sanford Dam, and everything including liability would revert to the Authority. Representative Peter DeFazio probably spoke for several when he questioned just how much money the taxpayers would be getting back for their investment in some of the projects being considered for relinquishment.[24]

In the case of the Canadian River, taxpayers got less than the full cost of the project. Recall that through reduced water deliveries and the drought relief act of 1996, the Authority's members had avoided paying some of the interest due on the cost of the facilities. E. R. Moore, president of CRMWA's board of directors at the time, told historian Rusty Hawkins in an interview how John Williams managed to get the final bill reduced yet again. Once Congress approved the early payment for the project, Williams looked at the originally promised amount of water and argued quite persuasively that, as Hawkins phrased it, "the Canadian River Municipal Water Authority had been misled by the original figures, and should have its repayment based on real figures" instead of the

Army Corps of Engineers' 1949 calculations and forecasts. So instead of roughly $50 million, CRMWA's members would pay the aforementioned $34,806,731. If some taxpayers, like Representative DeFazio, complained about the $15 million discount, it is probable that John Williams, E. R. Moore, former representative George Finger of Borger, and other current and past CRMWA members and supporters would have argued that the Authority was still paying $15 million out of "their own pockets" for water and not burdening state or federal taxpayers with the groundwater project. As it was, on March 25, 1999, CRMWA paid the Bureau a combination of CRMWA revenue bonds and cash (given by the City of Lubbock) and received title to the entire system. Local interests had overridden federal ownership and eleven cities now possessed the Canadian River Project, dam, pump, and pipeline.[25]

As the Authority argued with the Bureau, another Canadian River–based lawsuit entered the court system downstream of Sanford Dam. The case began in Roberts County and moved upstream as the State of Texas and landowners along the Canadian argued over where, exactly, the banks of the Canadian were and whether the dam's effects on the wet river had any effect on the banks of the paper river. To get an idea of just what was going on, imagine that you are driving north on Texas Highway 207 out of Borger. You go past a roundabout and the road drops abruptly a hundred feet or so into a steep-walled valley several miles wide. Drive beyond a pumping and transfer station and several clumps of cottonwood trees before rolling onto a long bridge. Looking over the side of the concrete bridge, you notice large amounts of brush but no water. Only at the very middle of the span, perhaps, do you see a small stream winding between the brush and grass; then you are back over sandy, overgrown land before the bridge ends and you see a roadside picnic area with cottonwood trees and a historical marker informing you about the lost community of Plemons. Ahead, the road climbs a steep slope leading out of the Canadian's inner valley. Obviously, the river had once flowed over much more of the valley floor, and if you were to get out of your car, duck under some fences (mind the snakes!), and look carefully, you would be able to find the remnants of some of the pre-dam riverbanks. The edge of the wet river had moved and logically, the edge of the paper river should have moved with it. According to the Texas General Land Office (GLO), that was not the case.[26]

Oil royalty rights and off-road vehicles led to the lawsuit. In a brief history of the dispute, attorney Jody Sheets pushed the date of the disagreement back to shortly after WWII, when the use of four-wheel-drive vehicles such as surplus Jeeps became popular with hunters. Some of the hunters drove along the Canadian's bed, ignoring property lines and fences in their pursuit of deer. When the ranchers filed trespass complaints against the hunters with the Hutchinson County justice of the peace, they learned that the state had cautioned the JP and local attorneys not to try prosecuting the violators because the state remained uncertain as to exactly what was state riverbed land and what belonged to private owners. This uncertainty also applied to who should get the royalties and lease payments for oil and gas wells drilled in the former riverbed. The Huber Oil Company of Borger had been working with exploration and leases in the area since the 1930s and noted that no gradient boundary survey of the Canadian had ever been made. This became important in the 1980s when the price for oil and gas increased, stimulating local exploration and drilling. The GLO, according to correspondence found during the discovery period of *Brainard et al.*, began pushing Huber to drill its state land leases. Huber's owners hesitated because ground surveys suggested that the old riverbed leases now sat on private land instead of on the state's riverbed. In 1987 the General Land Office ordered a new survey of the Canadian River in order to determine exactly where the banks were. The results, and the reason given for the results, drove E. H. Brainard II, Carolyn Rogers, Frances W. Klein, the Morrison Cattle Company, Boone Pickens, and others to hire their own surveyor and to sue the General Land Office.[27]

The landowners strongly disagreed with the results of the state survey. Darrell D. Shine, a surveyor hired by the GLO, studied archival records and aerial photographs and did a bank survey of the valley in Roberts and Hutchinson Counties. He used the "gradient boundary" method of locating the water line and then the bank. According to *Oklahoma v. Texas*, the boundary is "on and along the bank at the average or mean level attained by waters in the periods when they reach and wash the bank without overflowing it," or what hydrologists call the point of bank-full discharge. The state held that the historic banks and boundary gradient predating the closure of Sanford Dam remained the official and legal banks of the river. Sanford Dam, an "artificial" construction and obstruction to the river, had altered the wet river, but because the dam was manmade, its effects did not affect where the state's land began. Therefore, the legal banks of the river remained the pre-1965 banks, wherever they were.[28]

By using this argument, the state followed the "artificial-change theory" developed for dealing with human alterations to ocean beaches. Under state law, sand that accumulated and formed dry land without the intervention of a landowner belonged to the landowner—she now had more beach. If she hired a contractor to add sand in order to form a beach, it constituted an "artificial change" and so the new, manmade beach belonged to the state just as if the area were still below the water-line. In the case of the Canadian, this meant that any improvements, fences, and oil wells built after 1965 in what had once been the wet river's bed remained on state land and so taxes and royalties went to the GLO. And hunters and ORV users were not trespassing, because they traveled on state-owned property.[29]

E. H. Brainard, the Whittenberg family, and others disagreed with this decision for several reasons. A look at the original plat maps for Hutchinson and Roberts Counties (and Potter and much of Oldham as well) shows that surveyors divided the Canadian Valley in "long lots," long, narrow strips of land, so that more people could have access to the river. Long lots originated in Europe in order to give as many communities or individuals as possible access to various resources, such as waterfront, pasturage, and woodland. These New World ranchers and others owned riparian land and by law should have had access to the river or stream along their property. If the river channel shrank, then their property lines extended onto the new land so as to maintain contact with the all-important river at the gradient boundary. That humans rather than, say, beavers or a landslide built the dam constricting the channel did not matter, or so Brainard and others argued. Their surveyor, using the 1987 river channel as a guide, produced a rather different map of the river channel and the congruent property lines. They, the landowners, did not build the dam or take any actions on their own to expand their property into the river. It was the river that shrank, from the 3,400 feet (1,037 m) width the state claimed to the 20–50 feet (6–15 m) the private surveyor marked.[30]

The case began in the 1980s but did not finish winding its way through the Texas legal system until 1999, as the state tried very hard to have the old riverbed declared the official riverbed. The case first came to court in Roberts County (31st Judicial District) in 1989, and Justice Kent Simms held that "the riparian owners are entitled to have their land abut and be washed by the present flow of water, as established by . . . gradient boundary survey done under present conditions of the Canadian River and not as of a date before Sanford Dam closed."[31]

Understandably unhappy with this verdict, the State of Texas requested and received a change of venue to Hutchinson County. Here a local sportsman's group petitioned the court to come in on the side of the state. However, this court also ruled against the General Land Office, holding the Shine survey inadmissible and striking the testimony of well-known hydrologists Stanley A. Schumm and James Estes in the process. The landowners requested compensation for their costs, leading to another change of venue, this time to Collingsworth County. A jury in Wellington, the county seat, found in favor of the landowners and awarded the group $1,400 for surveyor's fees, $3,500 in attorneys' fees for trial, and $15,000 in fees for the appeal to the court of appeals. To no one's surprise, the GLO petitioned the state appeals court on the grounds that first, the trial court erred in accepting the landowners' survey because it was "premised on incorrect legal principles"; second, there were questions of fact concerning the correctness of the survey; and third, the court erred in not admitting the state's expert witnesses' testimonies and in denying the state's initial and subsequent motions for summary judgments based on the state's assertions that D. D. Shine's survey remained the correct one. The state also argued that the GLO should not have to pay the plaintiffs' fees. The landowners filed counterclaims.[32]

The appeals court ruled in favor of the GLO on all counts but ignored the artificial-change question. To the landowners' dismay, the court held that because Shine had used current as well as historic data, his survey was, indeed, a "present-day gradient survey." Also, the landowners did not have a right to have their land "washed by the present flow of water" because, as Col. Arthur Stiles wrote in the seminal study of gradient boundary determination, the "gradient boundary is determined by the bank of the river, not by the water in the river." Furthermore, the state was justified in getting a change of venue from Roberts County, it did not owe the landowners any fees, and the landowners were actually at trespass in an attempt to gain title to more land at the expense of the state. The plaintiffs and defendants all filed for review.[33]

Twelve years after the GLO authorized the state's survey, the Texas Supreme Court heard the arguments on February 10, 1999. The attorney representing the Canadian River Valley landowners, Michael V. Powell, outlined his clients' arguments quite simply. "First, our clients' lands were patented out of the State of Texas as riparian lands. They are, and are entitled to remain adjacent to the river, which is the essence and very most valuable feature of riparianness [sic]. Second, what the state owns is the bed of the river." Continuing, Powell explained that the artificial-change theory did not apply—the cause of the change was irrelevant. What mattered to the landowners and the state was that the river had changed greatly between 1965 and 1987, and that riparian property law meant that the banks of the Canadian River as it currently existed determined the edge of the Brainard's property. If a flood changed the river, then the riparian boundaries changed with it. However, flood precedents did not apply to the current situation. The attorney also pointed out that it was oil-rights questions that led to the original precedent in *Oklahoma v. Texas*. Chief Justice Nathan L. Hecht asked about the area immediately downstream of the dam, and Powell said that that area was different from his clients' land, in part because of the flow of water, and admitted, "I don't know what one would do if you were to set out to make a gradient boundary survey up there. But there is flowing water. It runs in small rivulus [*sic*]." Powell concluded that the GLO's gradient boundary survey was not proper, and affirmed to Chief Justice Hecht that he had evidence to prove it. [34]

The state also wanted "two critical questions answered" according to state attorney Mary A. Keeney:

> The first is whether the state is correct in its contention that the accretion bank [*sic*] formed by the sediment deposits from the water flow of the Canadian river constitute the proper banks on which to establish the gradient boundary. The second, is whether the state is correct in its assertion that the decrease in the river's water flow resulting from the closing of Sanford Dam in 1965 cannot be used as a basis for defeating the state's ownership of its riverbed.

Keeney said that the state accepted the gradient boundary determination, "until we got to the point of the flowing water." The bank, according to the state, did not have to touch flowing water, but according to their expert witnesses and one surveyor, when there was rainfall of more than one and a half inches, water did indeed reach the old 1965 riverbanks. The braided channel, caused by the inflow of eighteen tributaries, plus the 8 percent of the Canadian's flow that still passed Sanford Dam, extended wider than the private surveyor suggested and proved that the wet river still made use of the old bed. Keeney claimed that, according to Stiles's surveys, the tributaries "basically feed water into the bed, and put water in that bed basically every time it rains." This mattered because intermittent streams still counted as state-owned streams if they carried water after each rainfall. Keeney noted that this was the "State's riverbed" that they fed, not necessarily the current channel. Justice Deborah Hankinson expressed some confusion as to why the state used thirty-year-old data in its survey, to which Keeney replied that the dam did not matter. It cut off the water but did not change the state's riverbed. Chief Justice Hecht pushed that point, saying, "But if the river just dried up of its own, would you still own the bed?" Keeney replied, "If the river completely, naturally abandoned this area, I think the state *but for the minerals* would lose its interest in this bed" (emphasis added).[35]

The state maintained its claims to the pre-dam riverbed because of the minerals, in this case oil. Since Sanford Dam formed an artificial change, the old boundary gradient, although untouched by water, remained the correct edge of the riverbed. The current channel did not matter, and neither did traditions of riparian land ownership. In response to a question from Justice Harriett O'Neal, Keeney admitted that "we don't get enough inundation to scour all of that hydrophilic floor [plants] that we've got growing in that soggy sand bed that still exists today. . . . And so what [the landowners] have used is the absence of that kind of inundation and they have gone out and found a thirty foot channel." Under rather pointed questions from Justice Owen, the state's attorney admitted that no evidence existed "one way or the other" that water ever touched the 7,000-foot bank claimed by the state. The thirty-foot channel came about because of Borger's treated sewage and some industrial discharge that cut the narrower channel in the larger bed.[36]

The arguments in the case can be summarized as follows. The Canadian River shrank between 1873 (the initial survey date) and 1989. According to the State of Texas, the river's banks did not change even though the river no longer reached those banks, and brush and even trees now grew between the banks and the active river channel. The reason the banks remained the same in 1873 and 1989 stemmed from the fact that humans built Sanford Dam, causing an "artificial change" that altered the flowing river but not the legal edge-of-water. The landowners maintained that since they had bought the property in order to have contact with the river, when the river moved so too did their property lines. Since none of them built Sanford Dam with the intention of moving the riverbanks, they should not be penalized for taking advantage of the results of the dam's construction. In short, the Texas Supreme Court had to decide two things: first, if a waterless river still had banks and if so, where were they; and second, if a legal difference existed between Sanford Dam and a beaver dam.

It is worth keeping in mind what made this case so different from many disputes over riverbed use: the lack of federal land. Unlike the majority of river control projects in the American West, no federal land existed in the Canadian River watershed in Texas until the federal government bought the land under and upstream of Sanford Dam. Had the river flowed through public lands, as so many western rivers do, any question of riverbanks and royalties belonged to the Bureau of Land Management or the U.S. Forest Service or Park Service and federal law. Instead, only the State of Texas and state citizens had any interest in locating the exact edges of the somewhat-waterless Canadian River. Like so many parts of the river story, the lack of federal land both complicated and simplified the royalty and riverbank question. Any oil exploration within the riverbed would have to meet federal water and environmental protection regulations, but drilling rights and lease payments fell strictly under state and county law. The federal government watched the case but remained outside of it, something unusual to the point of near uniqueness in western river history.

The verdict the court issued in October 1999 failed to entirely satisfy the parties involved, much like the U.S. Supreme Court decision described earlier. The Texas Supreme Court ruled the state's use of the artificial-change theory invalid. It did not matter what caused the Canadian River to (in effect) stop flowing upstream of the plaintiffs' property. The fact that the plaintiffs had not gone out and narrowed the river channel and bed themselves for their own benefit made the fact that humans instead of "nature" built Sanford Dam irrelevant. The riverbed had narrowed, and riparian property doctrine allowed Brainard, the Morrison Ranch, and other property owners to take possession of the new land. "We reject the State's claim that the artificial change theory, as developed in these abandonment cases, dictates that the ownership of the riverbed, as it existed before the operation of Sanford Dam, cannot be altered," the court stated. Furthermore, "Allowing a riparian owner to gain title to new land formed by accretion or reliction that was influenced by artificial means comports with the policy rationales underlying the traditional doctrines of riparian ownership." The court also threw out the GLO's survey. However, the plaintiffs were not entitled to any financial awards because the GLO's actions were not "unreasonable."[37]

In sum, the Texas Supreme Court decided that the banks of the wet river and the banks of the paper river remained identical. The Canadian, classified as a navigable river because it was at least thirty feet (nine meters) across at its mouth, remained property of the state, all thirty or forty feet of it, while the former river bed now belonged to the landowners. The entire argument no doubt sounded foolish in the extreme to anyone living where rivers look like rivers, carry water year around, and are truly navigable by canoe, inner tube, or boat. After all, you cannot have a riverbank without having a river. Except that rivers move, a source of vexation going back to the first time any human tried to establish a fixed boundary along the bank or bed of live water. Although the General Land Office and the disappointed hunters and four-wheeler riders growled and grumbled, local interests pushed against the state and won, at least for the moment.[38]

And none of it would have come about if the black gold under the ground had not mattered so much. Again, as in the 1920s and on the South Plains in the 1950s, oil and gas development and revenue pushed people to deal with the Canadian River. In the 1920s the Borger oil boom made the Panhandle population soar and sent Amarillo and Borger scrambling for water while the oil companies tapped the blue gold of groundwater as well as the black. By the 1950s, the southern tip of the Ogallala, never as water rich as the thicker northern parts of the formation, succumbed to drought, municipal well use, and oil-field pumping, leaving Tahoka and other towns in dire straits. Now the hunt for oil royalties and for new sources of state revenue had led the court system to confirm what common sense would seem to have long since decided: that the river banks moved along with the river, dam or no dam. In essence the Texas Supreme Court put the river back where it had always been, but only because of oil.

And so, at the end of the last decade of the twentieth century the Canadian River returned to its original course. Or rather, the wet river would return to its original course if the Canadian River Municipal Water Authority ever shifted completely from river water to groundwater, something that it seemed increasingly likely to do over time. Groundwater would return to the Canadian that most people used, but now it would come from downstream of the reservoir via pumps and would flow out of faucets, not along a sandy bed. The U.S. Supreme Court had ruled that water in a reservoir all counted toward the amount of water in a reservoir, more or less, and that the same drop of water could originate in two or more places within a state, if that state was New Mexico. Nine years later, the State of Texas learned that the legal Canadian and the wet Canadian shared the same bed and banks downstream of Sanford Dam. All in all, it was a strange decade in the "life" of the river.

In the end, the 1990s marked the years when local interests took control of the fate of the Canadian River in the Texas Panhandle, at least for the near future. Eleven thirsty cities bought their water system back from the Bureau of Reclamation more than twenty years earlier than planned and for less than they had originally agreed to pay for it. To the disappointment of hunters and other riverbed users, riparian landowners in Hutchinson and Roberts Counties downstream of Sanford Dam persuaded the Texas Supreme Court that riparian rights and thirty years of dam-induced channel narrowing tied the wet and paper rivers to the same physical channel in the sandy bed at the heart of the Breaks. No one doubted that the state and federal governments would remain important to the life of the river and of the people living around it. But bottom-up demands and desires by at least some of the local people pushed major riverine developments in the 1990s.

A Once and Future River

I f the story of the Texas reaches of the Canadian River had stopped in 1973, the story could well have concluded with those famous words, "happily ever after." One might have imagined A. A. Meredith's ghost smiling at the sight of the almost brimful reservoir and stout dam straddling the Canadian's cliff-faced valley. His vision had come true, and eleven cities and towns made full use of the river's formerly "wasted" waters. A new playground for fishers, hunters, and boaters glistened in the full moon's light where once only bison and cattle had trod. But the story did not stop there, and new generations with changing ideas about conservation and the role of people in their landscape altered how some thought about the Canadian, while a combination of human action and shifting weather slowly dried up the mighty lake. But still, despite everything, the Canadian trickled and occasionally surged along its bed and through the Breaks.

The story of the Canadian River is one of goals accomplished, of the reassertion of local authority over the interests of federal and state agencies, of the influence of the petrochemical industry, and of the limits of human technology and planning. River authorities conserved the Canadian River between Ute Dam and Sanford Dam, just as a cook conserves a leg of lamb by serving it as roast lamb on Sunday, lamb hash on Monday, and lamb stew on Tuesday, while making lamb stock out of the bone and other bits. Not a drop of water that could be reached with a pump went unused, either for swimming and boating, or for drinking and watering lawns, or for cultivating crappie, bass, and fish stories. As Gifford Pinchot had stated at the beginning of the twentieth century, managers put the river's waters to use for "the greatest good, to the greatest number [of people] for the longest time," if that greatest good was providing drinking water and green lawns to urban residents. Many more people had tasted, floated, or stalked the river's waters than had before 1965, thanks to the recreation area and pipelines. A. S. Stinnett, John McCarty, and A. A. Meredith would have nodded in approval, while Gene Howe's ghost probably appreciated the fishing opportunities.[1]

Local interests and politics drove the development of the Canadian River in the Texas Panhandle. The municipal leaders of Amarillo and Borger, joined by those of Lubbock, Tahoka, O'Donnell, Lamesa, and other towns, agitated, planned, lobbied, and collected funds in order to persuade the federal government, via the Bureau of Reclamation, to build a dam and pipeline in order to provide water for drinking and to protect regional irrigation interests from municipal water development. When the surface waters ran dry, those same cities set aside their differences and once again lobbied, wheedled, and eventually forced the Bureau to accept early repayment for that same dam and pipeline so that CRMWA could pump groundwater from the Canadian watershed south to make up the lack. While the federal government still regulated water-quality standards and supervised the recreational area and Alibates Flint Quarries National Monument, local interests ran the river.

The regional petroleum industry pushed and pulled those local interests to a certain extent. A. S. Stinnett did not know about the Dixon Creek oil field when he helped pay for a railroad bridge across the Canadian in 1923, but oil soon rivaled agriculture in the regional economy and politics. The petro-born population surge in the 1920s and 1940s gave the region more political clout in Austin and Washington. The quest for black gold led to the creation of Borger and fed Amarillo's growth, pushing the towns to look for more resources in order to support their new populations. Although not completely dependent on oil, natural gas, and their related industries for survival, the Panhandle and South Plains embraced the sticky, smelly fluid. Developing oil required more water to support the equipment as well as the personnel, and the two industries shared technology and frustrations when drill bits yielded only dry holes. To an extent, the petroleum economy helped local water developers regain control over "their river" from the federal government.

This assertion of local control seems to be the opposite of the theory of hydraulic government popularized by historians Karl August Wittfogel and Donald Worster. Instead of a top-down imposition of federal management, local groups and agencies used the federal government as a resource before asserting final control over the Canadian. It is quite possible that the federal government granted this control simply due to a lack of interest in the minor stream in a relatively poor and unpopulated area, making local dominance illusory rather than actual. And federal agencies remained active in the river valley, through grants for phreatophyte control to CRMWA and the Fish and Wildlife Department's investigations into a potentially endangered species of shiner. That said, the fact remains that the Bureau sold the Canadian project back to CRMWA and that the Texas Supreme Court supported local riparian property owners in their argument with the state over property borders and riparian rights.

Drawing back to look at the Canadian River story in the larger national picture, it would be easy to point to it as yet another example of Anglo American, Western, or even human "destruction of the environment." People dammed and drained a once free-flowing river in order to support an unsustainable lifestyle and culture. The pristine Canadian dwindled into a creek flowing through a mudflat and puddle. As with the Colorado in Arizona and California, or the Arkansas in the High Plains of Kansas and Colorado, the wild Canadian River vanished, a victim of overuse and domestication. It is the fall from Eden told through hydrology. Except it is more than that, much more.

The first question posed to those who decry the loss of the "pristine" river should be: which river? The Canadian, like other plains streams, remained highly variable, cutting a channel or building layers of sediment, dry in some years and ripping out banks and trees and bridges the next. If one wished to see how the Canadian appeared prior to the permanent arrival of Euro Americans, perhaps in the days when the Comanches dominated the region, she might find an overgrazed, partly deforested stream with deeply eroded tributaries cutting gullies into the Breaks. That is, if she happened to look at the valley in the early 1860s as a decade of drought gripped the High Plains and Rio Grande Valley. The "paradise" described by Ynocencio Romero and surveyor W. S. Mabry in 1876 and 1873 respectively owed some of its lushness to the local extinction of the bison and some to a series of wet years. The combination allowed the grasses and other vegetation to recuperate, reducing sediment runoff and turning the Canadian into an eroding, or "degrading" stream. Come the drought of the 1880s, the pattern again reversed. Grazers and shifting rainfall patterns pushed the Canadian farther toward agradation as it rebuilt the eroded channel, spreading over the cow-cut banks. Floods in 1893, 1904, 1935, 1941, 1955, and 1967 affected the Canadian Valley just as they always had, washing out the banks, depositing or removing sediment, widening the channel, and leaving fresh sand or silt as the waters retreated. The Canadian had never been a static place, unchanging or constant.

The desire to return to an "unspoiled" environment runs through Western culture and some branches of science. A longing for a return to a lost golden age—be it a Judeo-Christian Eden, the reign of Numa Pompilius, or the time of the three great innovators in ancient China—runs deep in many human traditions. Some sciences also focus on the prehuman state of an environment and occasionally envision that as the preferred state. The ecologist Robert V. O'Neill addresses this in his essay, "Is It Time to Bury the Ecosystem Concept? (With Military Honors, of Course!)" He points out that the traditional idea of closed ecosystems, most famously the textbook example of a pond, no longer matches what different branches of science and history show. The concept, once useful, has become limiting, and perhaps it is time to look for another idea. O'Neill reminds his readers that there never was a "fall" from a "perfect" environment: humans are a keystone species, but other species also play critical roles in shaping the physical environment. The botanist Frederic Clements's idea of a final, steady-state "climax" environment of plants, animals, insects, and climate remained incomplete, as botanists and other biologists well know. O'Neill highlights the importance of chaos and change in how humans study their surroundings. This applies to the Canadian River as much as to the legendary ecosystem pond.[2]

The Canadian in 2014 is not what it was in 1800, or in 1950 for that matter. But it has never been stable, at least not that fluvial geomorphologists have been able to determine. The series of dams that blocked the river in the twentieth century changed the river but did not completely domesticate it, in part because of the unintended consequences stemming from their construction: changes in erosion and in the botanical composition of the riparian vegetation, and the ways the reservoirs would affect groundwater around and downstream of the dams. Much remains unknown about the current and future hydrology and culture of the Canadian watershed as well: the effects of ongoing changes in the weather patterns, the results of shifting land uses and groundwater appropriation, the effect of new technologies and economies of water use on the Canadian and the Ogallala Aquifer. Over time, rivers vanish and reappear or relocate as climates change and the land beneath the river sinks or rises, erodes or uplifts. These processes happened long before humans began channeling streams for their own purposes. It is possible that a series of wet, stormy years such as the 1960s and 1970s could refill the reservoirs, or another 1941 could send the Canadian raging past Sanford Dam's spillway. To paraphrase the Monty Python's "Dead Parrot Sketch," the river is not completely dead—it is "just sleeping."

The greatest change in the hydrologic and benthic nature of the Canadian is that it now consists of three rivers. One flows from the headwaters to Ute Dam. The second stream begins at the base of Ute Dam and continues to Sanford Dam, while the final river leaves Texas and eventually reaches the sea. There were times in the past when the river went dry in stretches, temporarily stranding fish populations or eliminating them from sections of the river's course. Now the Canadian's waters form three discrete sections. Water managers for CRMWA consider the effective watershed of their river to include only the area from Ute Creek east. There is simply not enough water left to flow out of Sanford Dam's spillway to connect downstream, aside from the seepage. In this instance, the alteration to the river appears to be unique, at least given the currently available geologic and hydrologic information. But another Canadian River still runs freely from New Mexico to the sea, although it travels by way of Austin and Santa Fe.

While the wet river dwindled, at least for the foreseeable future, the paper river flowed briskly through courtrooms and newspaper offices. Neatly divided between states and carefully defined by lines on charts and maps, the paper river began in Santa Fe, washed up in Austin, then continued on to Oklahoma and out of the state's purview. The Canadian River Municipal Water Authority, CRMWA, owned the flow of the paper river in fee simple with the blessings of the state. The State of Texas claimed the sand below the stream and the oil, gas, gold, or other minerals below those sands while allowing members of the public to hunt and fish and drive along the state lands. On paper, 150,000 acre feet of water per year went to CRMWA, and the rest remained free for the state and state residents to claim, within the limits of the Canadian River Compact and Texas law. The fact that the wet river failed to abide by these agreements and could not be neatly sorted like sausage slices provided employment for water lawyers and engineers alike during the mid-twentieth century. The mystery of groundwater returned to Austin for the courts to solve, and they relegated the task to the state legislature, which remained reluctant to take on the Gordian knot of pumping limitation. High Plains residents desired legal resolution but at the same time worried about the decisions of a legislature from what sometimes appeared to be a totally different state.

One minor but intriguing thread winding through the history of the Canadian in both Texas and New Mexico is that of local and regional identity. Prior to 1850, the watershed formed a single entity for Hispanos and Comanches, despite lines drawn on paper in 1836 and again in 1848.

After the creation of New Mexico Territory, the differences in land management and sale by the State of Texas and federal government contributed to the differing settlement patterns of the river valley, even as Hispanos tried to maintain the cultural unity of the river. As settlement in both states reached the High Plains, a similar extractive economy based on agriculture developed, closely followed by the petroleum economy, then water mining where groundwater was available. Residents of Texas assisted New Mexicans with obtaining Conchas Dam, and New Mexicans took part in the Panhandle Water Conservation Association during the 1930s. The divisions of the 1950s–90s at the state level did not stop residents of Texas from putting boats into Ute Lake or keep Tucumcari citizens from shopping in Amarillo. CRMWA worked with the New Mexico State Engineer's Office and other agencies in its efforts to control salt cedar. The grumbles of Lubbock residents about deafness in the capitol at Austin drew sympathetic nods from Claytonites irked at the latest decrees from Santa Fe. Everyone groused about Washington, D.C., even as they took federal moneys for crop insurance, infrastructure projects, and other benefits. A lingering watershed mindset flowed beneath the surface of the High Plains much as the Canadian flowed through the lowlands.[3]

This cultural unity stemmed from several causes, some of which have been touched upon in this book. The dominant Anglo American traditions of resource utility and growing commoditization of water, soil, and grass play familiar roles in the river story. A shared semiarid climate that limits economic opportunities in agriculture as well as capping the region's faunal populations also shaped local culture. The "lords of yesterday," old ideas of economic value and water use, bind the region to an extraction and acquisition pattern rather than a modern conservation mindset, at least for the moment. The watershed's legacy of "Old West" popular culture, as seen in local rodeos, musicals, popular historical interpretation, and resistance to some outside ideas, provides stability but also provokes conflicts with new residents coming from outside the Anglo American cultural sphere and occasionally collides with state and federal policies.[4]

The tension between wanting federal help and resenting the government that accompanied it emerges in the Canadian's story after 1925. Westerners in general and Texans in particular value the ideas of local independence and private property rights. Panhandle water dreamers needed federal support in order to obtain their goal, but as time passed they grew more and more reluctant to pay the price for that help, either in cash or in what they saw as the loss of their rights to do as they wished. Because it lacks federal land, Texas faced this dilemma later than did states such as Wyoming or Nevada. However, Texans reacted the same way that other westerners did, except that in this case they succeeded in buying out the Bureau of Reclamation. The argument over federal benefits and rights versus local and private gains and resource continues, whether it be over oil drilling leases on Bureau of Land Management or Forest Service land, or the regulation of fishing within U.S. coastal waters, the proper price and management of grazing land, or the distribution and control of waters west of the 100th meridian.

Still the Canadian and the Breaks remain a physical presence, a faint blue line on the horizon of the Llano Estacado. Despite a century and more of ranching and irrigation, deer and cougar make the Breaks their home. Rumors of elk and black bear draw curious speculation, while beaver have returned to their old habitat, at least temporarily. A local thunderstorm that dumps enough rain and hail in the right place sends the river rising up to a foot in a matter of hours. Local historians including Delbert Trew, John Erickson, and others preserve the memories of how ferocious the Canadian once was and might be again. The Canadian is a river domesticated but not yet completely tamed, one that shows the possibilities of cooperative action and the pitfalls of unintended consequences. The irrigation that CRMWA members sought to protect in turn drained water from the red waters that they intended to substitute for groundwater's blue gold. Plants introduced to prevent erosion took over large stretches of the Canadian and its sister stream, the Pecos River. Climatic changes tossed a wild card into the water game by increasing evaporation and reducing snow and rainfall, then reversing perhaps.

Yet the vermillion-colored river still runs. It trickles through the valley, sloshes through courthouses and legal offices, and drips from water taps, part of the great water story of human history and a source of pleasure, sorrow, wonder, and frustration for those who live within the old and new Canadian watersheds.

ABBREVIATIONS, ACRONYMS, AND TERMS

Abbreviations
ppm: parts per million.
a/f: acre-feet—enough water to cover one acre of ground in one foot of water; one acre-foot equals 325,851 US gallons.
lps: liters per second.
mg/l: milligrams per liter—amount of a chemical per liter of water.
AUM: animal unit month—a measurement of grazing capacity.
GPM: gallons per minute.
GPD: gallons per day—describes production of a well or municipal usage.
CFS: cubic feet per second—a measure of stream flow; how much water goes past a certain point in a certain period of time.
k/gal: per one thousand gallons—cost description of water.

Acronyms in Order of Appearance
USDA: United States Department of Agriculture.
TBWE: Texas Board of Water Engineers. After 1965 became the Texas Water Rights Commission (TWRC). *See also* TWDB below.
TSDH: Texas State Department of Health. After 1975 it became the Texas Department of Health.
AAA: Agricultural Adjustment Act.
WPA: Works Progress Administration.
PWCA: Panhandle Water Conservation Association.
SVA: Southwestern Valley Association.
CRPOC: Canadian River Project Organizing Committee.
CRWUA: Canadian River Water Users Association.
CRMWUA: Canadian River Municipal Water Users Authority. CRWUA and CRMWUA may be the same organization. The two names are used at the same time and the difference may lie with newspaper reporters.
CRMWA: Canadian River Municipal Water Authority ("krim-wah"), 1953–present.

SPS: Southwestern Public Service, a utility company, now component of Xcel Energy.

RRWA: Red River Water Association.

TWDB: Texas Water Development Board.

ENSO: El Niño-Southern Oscillation, the macro weather pattern in the Pacific Ocean.

NPS: National Park Service, oversees Lake Meredith National Recreation Area and Alibates Flint Quarries National Monument.

TWQA: Texas Water Quality Administration.

TPWD: Texas Parks and Wildlife Department, state agency that oversees recreation, hunting, fishing, conservation, and habitat protection on state lands.

CRWA: Colorado River Water Authority, oversees the Colorado River in Texas, south of the Caprock, 1960–present.

EPA: Environmental Protection Agency, federal agency created by the Environmental Protection Act in 1969 to oversee implementation of environmental quality legislation and regulation such as the Clean Air Act.

ESA: Endangered Species Act, a law allowing the federal Fish and Wildlife Agency to develop plans and regulations to protect those plant and animal species believed to be in imminent danger of becoming extinct.

GLO: Texas General Land Office, the division of state government in charge of management of state lands and the income associated with them.

Technical Terms

Agrading: a stream that is depositing material and building up its bed.

Degrading: a stream that is cutting down into its bed and deepening the channel.

Groundwater: water trapped in rock formations under the surface of the ground. Such a formation is called an aquifer, such as the Ogallala Aquifer, the largest in the United States. The water-bearing Ogallala Formation extends from South Dakota to the southern Llano Estacado and southeastern New Mexico.

Prior Appropriation: the legal doctrine governing much of water law in the western United States. The first individual to put a stream's waters to beneficial use, however defined by the state, has the first or senior right to that water.

Riparian Law: water laws generally used by states located east of the 100th meridian. Stream waters may be used by those with streamside property, but they cannot interfere with the water uses of downstream property owners.

Stream: the hydrologic term for any flowing body of water. All rivers are streams, but not all streams are rivers.

Usufruct: the legal right to use the fruits of a resource but not to fully consume that resource. Riparian water rights are usufructory: a property owner may water cattle or dip water for domestic use but cannot commandeer the stream's entire flow or contaminate it.

Watershed: the area within which precipitation drains into a stream. A watershed divide marks the place where drainage shifts from one stream to another, such as the Canadian and the Prairie Dog Town Fork of the Red River or the Canadian and the Washita.

Conversions and Climatology

Conversions

1 inch = 2.54 cm = 25.4 mm
1 foot = .305 m
1 mile = 1.609 km
1 acre = .405 ha = 4047 m^2
1 section = 640 acres = 1 mi^2 = 259 ha
1 cubic foot per second (cfs) = .028 m^2/s = 28.32 l/s
1 acre-foot (a/f) = 325,851 gal (US) = 1,233,482 liters

Climatology

Analyses of the potential changes in precipitation were done using a modification of the Balling and Wells methodology; see Robert C. Balling Jr. and Stephen G. Wells. "Historical Rainfall Patterns and Arroyo Activity within the Zuni River Drainage Basin, New Mexico," *Annals of the Association of American Geographers* 80 (December 199): 603–617. Data sets for Amarillo, Dalhart, Dumas, Hereford, Hartley, and Vega, Texas, and Tucumcari, New Mexico, were obtained from the national Climate Date Research Center in February 2011.

These data sets were broken apart by water and calendar year, and counts taken of: total days with precipitation, precipitation of 0.25–0.49 inches, 0.50–0.74 inches, and greater than 0.75 inches. The precipitation days were then broken down by water year into winter (October 1–March 31) and summer (April 1–September 30).

The number of intense events (greater than 0.75 inches per day), number of winter events, and number of low-intensity events (less than 0.25 inches) were graphed against years, along with an average of all stations and the trends analyzed. Aside from El Niño–La Niña correlations, no changes in the overall pattern were found during the period of record reviewed.

The graphical results and raw data sets are available as Microsoft Excel files. Interested readers may send an e-mail message to the author at RedWaterBlackGold@gmail.com and the Excel files will be sent.

FOR FURTHER READING

The High Plains and South Plains are a place apart both geographically and in the writing of their history. Early chroniclers focused on the stories of the great ranches and legendary cowboys, giving far less attention to the surrounding land and peoples. J. Evetts Haley recorded the stories of the XIT Ranch and Charles Goodnight, among others, while Ernest Archambeau, John McCarty, and more recently Frederick W. Nolan devoted more attention to the Hispano settlement and Wild West town of Tascosa on the banks of the Canadian. The southern plains ranches such as those belonging to Colonel George W. Littlefield and others found historians of their own, but it was not until Frederick Rathjen's 1973 (which was revised in 1998) book, *The Texas Panhandle Frontier* that a regional history became available. True to its title, *The Texas Panhandle Frontier* stops with the arrival of Anglo Texan settlement and the closing of the frontier by barbed wire and railroad tracks. What some might consider a more literary approach to the Spanish and Anglo American exploration period can be found in John Miller Morris's work, *El Llano Estacado*. Miller Morris looks at the differing attitudes to semi-arid lands brought by the Spanish and the Anglo Americans, and how those shaped the groups' perceptions of the region.

Paul Carlson and Lawrence Graves have written histories of Amarillo and Lubbock respectively. Carlson's work, *Amarillo: The Story of a Western Town*, is readily available, while *A History of Lubbock*, Graves's older edited work, remains out of print and awaits a successor. Illustrated urban histories fill in some gaps for both communities. At the same time, less well-known local historians carefully collected the stories of the area's smaller towns and of their counties. These books tend to focus on churches, schools, anecdotes, and genealogies, but a careful reading and comparison of different town and county histories can provide nuggets of information that would be impossible to find elsewhere. The Canadian River, however, although running through many of the stories, remained without a specific study.

There has been one earlier history of the Canadian River: *Through Time and the Valley* by native son John R. Erickson. His anecdotal account of riding along the river in 1972 is a classic in its own right, a loving history in casual dress. He covers the area downstream of Sanford Dam, from old Plemons to the Oklahoma border, describing the ranches and towns along the way.

Aside from the literature of the region, *Red Water, Black Gold* is intended to contribute to environmental history. In many ways, the field of environmental history grew from studies of western rivers and the Dust Bowl. Donald Worster, one of the pioneers of the specialty, studied the causes of the Dust Bowl in his 1979 work, *Dust Bowl*. His subsequent volume on the Colorado River and water development in California, *Rivers of Empire*, exposes nonspecialists to the "hydraulic empire" thesis originally proposed by Karl Wittfogel and shows how government from above grew in conjunction with attempts to divert and manage the Colorado River's waters for the benefit of California farmers and urban residents. Worster raises questions about the cost of such dams and diversions to society, costs that go beyond dollars to include regional independence and perhaps even spiritual considerations. Norris Hundley's book *The Great Thirst* challenged Worster's theory, arguing instead that California's complex hydraulic developments were driven from below, not dictated from above. Journalist Marc Reisner penned the entertaining, if at times scathing, *Cadillac Desert*, which fleshes out the PBS television series of the same name. While covering the same stories as Worster does in *Rivers of Empire*, Reisner spends more time with the characters (of which there are plenty!) and ventures into the Great Plains, while Worster takes a more sober approach and restricts himself to the Colorado River watershed. Other western streams, including the Columbia, Rio Grande, and Arkansas, have their chroniclers; some take a more literary approach to their chosen stream (Paul Horgan's *The Great River* or John Graves's *Goodbye to a River*), and others are more traditional, such as Donald Littlefield's excellent study of the lower Rio Grande *(Conflicts on the Rio Grande)* or James E. Sherow's *Watering the Valley*, about the western Arkansas River, irrigation, and interstate water law.

Water policy and the Bureau of Reclamation have their own historians. Donald Pisani's three volumes on water development, *To Reclaim a Divided West: Water, Law and Public Policy, 1848–1902; Water, Land, and Law in the West, The Limits of Public Policy 1850–1920*; and *Water and American Government: The Reclamation Bureau, National Water Policy, and the West, 1902–1935*, provide lenses of varying power for looking at the often tangled, confusing, but fascinating development of the federal government's involvement with developing water resources west of the 100th meridian. Charles F. Wilkinson looks at the recent past and future in his collection of linked essays, *Crossing the Next Meridian: Land, Water and the Future of the West*. He takes his title from Wallace Stegner's classic account of John Wesley Powell and the Colorado River, *Beyond the Hundredth Meridian*.

The Ogallala Aquifer also has had its chroniclers. Donald E. Green's *Land of Underground Rain*, published in 1973, was then and still remains the best account of the rise of irrigation on the southern High Plains. John Opie's *Ogallala: Water for a Dry Land* draws heavily on Green's work, then expands to cover western Kansas and modern agribusiness on the High Plains. Opie looks farther than does Green, considering the implications of the increasingly sparse resources in the Ogallala Formation. The most readable book, and most wide ranging, on the Ogallala is *Ogallala Blue: Water and Life on the High Plains* by William Ashworth. Ashworth covers the entire breadth of the Ogallala, from the Nebraska Sandhills to the cotton fields of the South Plains.

Martin V. Melosi is the premier historian of water in the modern U.S. city. His volumes on sanitation, water supply, and sewage, and most recently on municipal water infrastructure development *(Precious Commodity: Providing Water for America's Cities)*, explore how towns and cities sought for and dealt with water for industry and private use. Be not dismayed by the length of some of his works. Melosi tells the surprisingly interesting story of how municipal and private corporations struggled to provide clean, safe, and abundant drinking water while disposing (or not) of the "used" product.

For those interested in the history of dams in the western United States, an intriguing study of dam technology and politics is *Building the Ultimate Dam: John S. Eastwood and the Control of Water in the West* by Donald C. Jackson. In addition to looking at policy and personalities, Jackson gives readers a useful primer on different types of dams. Engineering alone did not always determine what kind of structure would be built to hold back the waters.

Readers will find a full list of works consulted in the bibliography.

NOTES

PREFACE

1. "What Is Meant by the Term Drought?," <http://www.wrh.noaa.gov/fgz/science/drought.php?wfo=fgz> [Accessed 22 June 22, 2014].

2. Roderick Nash, *Wilderness and the American Mind* (4th ed.; New Haven: Yale University Press, 2001), xiii, 2, 6.

INTRODUCTION

1. P. S. Wisniewski and F. J. Pazzaglia, "Epirogenic Controls on Canadian River Incision and Landscape Evolution, Great Plains of Northern New Mexico," *Journal of Geology* 110 (July 2002): 439, 441, 452; William R. Muehlberger, Sally J. Muehlberger, and L. Greer Price, *High Plains of Northeastern New Mexico: A Guide to Geology and Culture* (Socorro: New Mexico Bureau of Geology and Mineral Resources of the new Mexico Institute of Mining and Technology, 2005),102, 14–15; and Paul N. Dolliver, *Cenozoic Evolution of the Canadian River Basin* (Waco, Tex.: Baylor University Press, 1984), 13, 27, 40, 70–71; Erwin J. Raisz, *Landforms of the United States* (6th ed., Cambridge, Mass.: Erwin J. Raisz, 1957).

2. Mark W. Presley, "Salt Deposition Systems: An Example from the Tubbs Formation," in *Geology and Geohydrology of the Palo Duro Basin, Texas Panhandle: A Report on the Progress of Nuclear Waste Soil Feasibility Studies 1979,* ed. Thomas C. Gustavson (Austin: University of Texas Press for Texas Bureau of Economic Geography, 1980), 24, 25; Thomas C. Gustavson, "Structural Control of Major Drainage Elements Surrounding the Southern High Plains," in *Geology and Geohydrology of the Palo Duro Basin, Texas Panhandle: A Report on the Progress of Nuclear Waste Soil Feasibility Studies, 1980* ed. Thomas C. Gustavson (Austin: University of Texas Press for Texas Bureau of Economic Geography, 1981), 176, 177; 12, 14, 102.

3. Author's observation 1981–2011.

4. C. C. Reeves Jr. and Judy A. Reeves, *The Ogallala Aquifer of the Southern High Plains*, vol. 1: *Geology* (Lubbock: Estacado Books, 1996), 55, 57, 58, 128, 233, 247, 271.

5. William Abert, *Expedition to the Southwest: An 1845 Reconnaissance of Colorado, New Mexico, Texas and Oklahoma* (Lincoln: University of Nebraska Press, 1999), 45, 49, 54–62; François des Montaignes, *The Plains*, ed. Nancy Alpert Mower and Don Russell (Norman: University of Oklahoma Press, 1972), 116–121.

6. Ernest R. Archambeau, "Panhandle Pioneer Settler Recalls Origin, Early Days of 'Old Tascosa,'" *Amarillo Times*, Feb. 28, 1946, 2; José Ynocencio Romero to Ernest R. Archambeau, "Spanish Sheepmen on the Canadian at Old Tascosa," *Panhandle Plains Historical Review* 19 (1946): 47; and Roy Riddle, "Casimero Romero as Benevolent Don in Brief Pastoral Era,"

Amarillo Times, Aug. 14, 1938. For a full account of the Canadian River prior to 1940, see John R. Erikson, *Through Time and the Valley* (Denton: University of North Texas Press, 1995), and Margaret A. Bickers, "Three Cultures, Four Hooves and One River" (Ph.D. diss., Kansas State University, 2010), from which most of the following is drawn.

7. Vance T. Holliday, *Paleoindian Geoarchaeology of the Southern High Plains* (Austin: University of Texas Press, 1997), 124,125; Christopher Ray Lintz, *Architecture and Community Variability within the Antelope Creek Phase of the Texas Panhandle* (Norman: University of Oklahoma Press, 1986), 30; and Paul H. Carlson, *Deep Time and the Texas High Plains: History and Geology* (Lubbock: Texas Tech University Press, 2005), 32–33.

8. Lintz, *Architecture and Community Variability*, 33, 214.

9. Ibid., 253, 239.

10. Ibid., 239, 253.

11. Tom McHugh, *Time of the Buffalo* (Lincoln: University of Nebraska Press, 1979), 13, 17, 43; Douglas B. Bamforth, *Ecology and Human Organization on the Great Plains* (New York: Plenum Press, 1988), vi, 30, 54, 108; Gary Clayton Anderson, *The Indian Southwest 1580–1830: Ethnogenesis and Reinvention* (Norman: University of Oklahoma Press, 1999), 139, 213; Pekka Hämäläinen, *The Comanche Empire* (New Haven, Conn.: Yale University Press, 2008), 18, 23; Ernest Wallace and E. Adamson Hoebel, *The Comanches: Lords of the South Plains* (Norman: University of Oklahoma Press, 1952), 11; Elizabeth A. H. John, *Storms Brewed in Other Men's Worlds: The Confrontation of Indians, Spanish, and French in the Southwest, 1540–1795* (2nd ed.; Norman: University of Oklahoma Press, 1996), xix; and Hämäläinen, *Comanche Empire*, 9, 65.

12. James E. Sherow, "Workings of the Geodialectic: High Plains Indians and Their Horses in the Region of the Arkansas River Valley, 1800–1870," in *A Sense of the American West*, ed. James E. Sherow (Albuquerque: University of New Mexico Press, 1998), 100; Dan Flores, "All the Pretty Horses: The Horse Trade and the Early American West, 1775–1825," *Montana: The Magazine of Western History* 58 (Summer 2008): 5, 11,14.

13. Dan Flores, "Bison Ecology and Bison Diplomacy Redux: Another Look at the Southern Plains from 1800–1850," in *The Natural West: Environmental History in the Great Plains and Rocky Mountains* (Norman: University of Oklahoma Press, 2001), 62, 68–69; Sherow, "Workings of the Geodialectic," 107; David W. Stahle and Malcom K. Cleveland, "Texas Drought History Reconstructed and Analyzed from 1698 and 1980," *Journal of Climate* 1 (January 1994): 64; Hämäläinen, *Comanche Empire*, 329, 360–361; Romero, "Spanish Sheepmen," 55.

14. James R. Brooks, *Captives and Cousins: Slavery, Kinship and Community in the Southwest Borderlands* (Chapel Hill: University of North Carolina Press, 2002), 66, 133; Richard L. Nostrand, *The Hispano Homeland* (Norman: University of Oklahoma Press, 1992), 78–81.

15. Romero, "Spanish Sheepmen," 61, 49, 55, 56–57; Archambeau, "Panhandle Pioneer," 2; Pauline Durrett Robinson and R. L. Robinson, *Cowman's Country: Fifty Frontier Ranches in the Texas Panhandle 1876–1887* (Amarillo, Tex.: Paramount Publishing Company, 1981), 37, 106, 115, 128,152.

16. Robinson and Robinson, *Cowman's Country*, 37; J. Evetts Haley, *Charles Goodnight: Cowman and Plainsman* (Norman: University of Oklahoma Press, 1949), 276, 280, 298, 305; Donald F. Schofield, *Indians, Cattle, Ships and Oil: The Story of W. M. D. Lee* (Austin: University of Texas Press, 1985), 59; and Dulcie Sullivan, *The LS Brand: The Story of a Texas Panhandle Ranch* (Austin: University of Texas Press, 1968), 39.

17. Paul Wallace Gates, *The History of Land Law Development* (Washington, D.C.: U.S. Government Printing Office, 1968; reprint, New York: Arno Press, 1979), 66, 501; E. Louise Peffer, *The Closing of the Public Domain: Disposal and Reservation Policies, 1900–1950* (New York: Arno Press, 1972), 25, 2, 78; Haley, *Charles Goodnight*, 278, 280, 302–303; Thomas Lloyd Miller, *The Public Lands of Texas 1519–1970* (Norman: University of Oklahoma Press, 1972), 31.

18. William M. Pearce, *The Matador Land and Cattle Company* (Norman: University of Oklahoma Press, 1964), 83; Schofield, *Indians, Cattle, Ships and Oil*, 56; J. Evetts Haley, *The*

XIT Ranch of Texas and the Early Days of the Llano Estacado (new ed.; Norman: University of Oklahoma Press, 1953), 71–72; and W. Turrentine Jackson, "British Interests in the Range Cattle Industry," in *When Grass Was King: Contributions to the Western Range Cattle Industry Study*, ed. Maurice Frink, W. Turrentine Jackson, and Agnes Wright Spring (Boulder: University of Colorado Press, 1956), 137, 143, 145, 146, 147, 216.

19. Paul H. Carlson, *Amarillo: The Story of a Western Town* (Lubbock: Texas Tech University Press, 2006), 20, 21, 22.

20. O. H. Loyd to J. E. Haley, Letter, June 30, 1926, Folder: Oldham County, Loyd, O. H., Manuscript Files (Panhandle-Plains Historical Museum Archives, Canyon, Texas; cited hereafter as PPHM.)

21. "Panhandle Real Estate Journal," n.p., n.d., in Voucher 4, "Advertising," George Tyng Statement of Accounts Nos. 47–49, folder Francklyn D 3a 3715, Box 6 of 32, Francklyn Land and Cattle Company, Manuscript Collection (PPHM); Gary L. Nall, "Panhandle Farming in the 'Golden Era' of American Agriculture," *Panhandle Plains Historical Review* 46 (1973): 88, 89.

22. Stanley W. Trimble and Alexandra C. Mendel, "The Cow as a Geomorphic Agent: A Critical Review," *Geomorphology* 13, no. 1 (1995): 243, 247–248; John R. Vallentine, *Grazing Management* (San Diego, Calif.: Academic Press, 1990), 70–71; Romero, "Sheepmen," 57; Dave Rosgen, *Applied River Morphology* (2nd ed.; Pagosa Springs, Colo.: Wildland Hydrology, 1996), 2–2, 8–12; Thomas Dunn and Luna B. Leopold, *Water in Environmental Planning* (New York: W. H. Freeman and Company, 1978), 258, 261, 510, 516; Daniel H. Mann and David J. Meltzer, "Millennial-Scale Dynamics of Valley Fills over the Past 12,000 14C Years in Northeastern New Mexico, USA," *GSA Bulletin* 119 November–December, 2007): 1446; Devine Ethredge, "The Role of the North American Monsoon in the Landscape Evolution of the Southwest United States" (master's thesis, University of New Mexico, 2000), 73 74; S. A. Schumm and R. W. Lichty, *Channel Widening and Flood-Plain Construction along Cimarron River in Southwestern Kansas* (Washington D.C: United States Government Printing Office, 1963), 79, 82, 86; New Mexico State Engineer's Office, *Preliminary Report on the Geology of the Ute Dam Site, Quay County, New Mexico*, vol. 3: *Canadian River Storage Sites Investigating Ute Reservoir* (Santa Fe: State Engineer's Office Technical Division, 1961), Folder 1001, Box 29, S/N 4346, State Engineer's Records (New Mexico State Record Center and Archive, Santa Fe, New Mexico; cited hereafter as NMSRCA), 34, 37; Timothy Kane, Fluvial Geomorphology Lecture, Kansas State University, Dec. 1, 2005; and Jeffrey A. VanLooy and Charles W. Martin, "Channel and Vegetation Change on the Cimarron River, Southwestern Kansas, 1953–2001," *Annals of the Association of American Geographers* 95, no. 4 (2005): 727, 736–737.

23. United States Weather Bureau, "Flood on the South Canadian River in Oklahoma and Indian Territory, October 1–4, 1904," *Monthly Weather Review* 32 (November 1904): 522; John L. McCarty, *Maverick Town: The Story of Old Tascosa* (Norman: University of Oklahoma Press, 1946), 251; and Pearce, *Matador Land and Cattle Company*, 153.

CHAPTER 1

1. "Parity" refers to the attempt to ensure farmers' purchasing power remained at a stable level relative to farm costs, using the "Golden Age" of 1909–14 as a base. R. Douglas Hurt, *Problems of Plenty: The American Farmer in the Twentieth Century* (Chicago: Ivan R. Dee, 2002), 41, 43, 45; Deborah Fitzgerald, *Every Farm a Factory: The Industrial Ideal in American Agriculture* (New Haven, Conn.: Yale University Press, 2003), 19; and Gary L. Nall, "Specialization and Expansion: Panhandle Farming in the 1920s," *Panhandle Plains Historical Review* 47 (1974): 47, 49.

2. Hurt, *Problems of Plenty*, 44; Nall, "Specialization and Expansion," 50.

3. Nall, "Specialization and Expansion," 58, 60.

4. Ibid., 51; Hurt, *Problems of Plenty*, 52, 54; Fitzgerald, *Every Farm a Factory*, 11.

5. Hurt, *Problems of Plenty*, 55. Yes, the Volstead of the Volstead Act, better known as Prohibition.

6. Ibid., 57, 60.

7. Ibid., 57, 60; R. Douglas Hurt, *American Agriculture: A Brief History* (Ames: Iowa State University Press, 1994), 267, 268.

8. Nall, "Specialization and Expansion," 52, 53.

9. Hurt, *American Agriculture*, 248; Hurt, *Problems of Plenty*, 49; Nall, "Specialization and Expansion," 59, 60, 65.

10. Nall, "Specialization and Expansion," 61, 62; Donald Abbe, Paul H. Carlson, and David J. Murrah, *Lubbock and the South Plains: An Illustrated History* (Chatsworth, Calif.: Windsor Publications, 1989), 38, 57; "Cotton Marketing," <http://www.oldandsold.com/articles04/textiles5.shtml> [Accessed May 15, 2014].

11. Nall, "Specialization and Expansion" 61; Abbe et al., *Lubbock and the South Plains*, 37–38, 39; Donald E. Green, *Land of the Underground Rain: Irrigation on the Texas High Plains, 1910–1970* (Austin: University of Texas Press, 1973), 120, 121.

12. Hurt, *Problems of Plenty*, 48; Nall, "Specialization and Expansion," 67; Donald Worster, *Dust Bowl: The Southern Plains in the 1930s* (25th anniversary ed.; New York: Oxford University Press, 2004), 92–93; Lucien Burnett to Woody Coffee, Oct. 20, 1962, Ms/Int: Burnett, Lucien (PPHM), 14, 15.

13. Paul Bonnifield, *The Dust Bowl: Man, Dirt and Depression* (Albuquerque: University of New Mexico Press, 1979), 57–58.

14. James C. Malin, *Winter Wheat in the Golden Belt of Kansas: A Study in Adaptation to Subhumid Geographical Environment* (1944; reprint, New York: Octagon Books, 1973), 239, 240, 245; Sarah Phillips, "FDR, Hoover and the New Rural Conservation, 1920–1932," in *FDR and the Environment*, ed. Henry L. Henderson and David B. Woolner (New York: Palgrave Macmillan, 2005), 119, 121.

15. Phillips, "New Rural Conservation," 117, 119, 122, 137.

16. N. D. Bartlett, "Discovery of the Panhandle Oil and Gas Field," *Panhandle Plains Historical Review* 12 (1939): 49; Carlson, *Amarillo*, 83, 84; B. Byron Price and Frederick W. Rathjen, *The Golden Spread: An Illustrated History of Amarillo and the Texas Panhandle* (Northridge, Calif.: Windsor Publications, 1986), 87.

17. Price and Rathjen, *Golden Spread*, 87–88; Bartlett, "Discovery," 51, 52; H. Allen Anderson, "Borger, TX," *The Handbook of Texas Online*, <http://www.tshaonline.org/handbook/online/articles/heb10> [Accessed May 15, 2014]; Panhandle Geological Society, *Oil and Gas Fields of the Texas and Oklahoma Panhandles* ([Amarillo]: Panhandle Geological Society, 1961), 9, 10, 11; Hutchinson County Historical Commission, *History of Hutchinson County, Texas* (Dallas: Taylor Press, 1980), 28, 29. One barrel is 42 U.S. gallons.

18. Arthur W. McCray and Frank W. Cole, *Oil Well Drilling Technology* (Norman: University of Oklahoma Press, 1959), 325, 326–327, 328; F. Stanley, *The Early Days of the Oil Industry in the Texas Panhandle* (Borger, Tex.: Hess Publishing Co., 1973), 73, 147.

19. Price and Rathjen, *Golden Spread*, 88; Bartlett, "Discovery," 50.

20. Hutchinson County Historical Commission, *History of Hutchinson County*, 31; Carbon black is still produced in Borger today, and the older plants are very easy to identify, a legacy of the early period of production.

21. Bob Lasley and Sallie Holt (comps.), *Dust Storms and Half-Dugouts: Tales from the Good Old Days in the Texas Panhandle* (Hickory, N.C.: Hometown Memories Publishing Company, 2009), 338, 339–340; Hutchinson County Historical Commission, *History of Hutchinson County*, 31, 290, 374.

22. Hutchinson County Historical Commission, *History of Hutchinson County*, 21, 371.

23. Hutchinson County Historical Commission, *History of Hutchinson County*, 21, 371, 443;

George Tyng to Francis De P. Foster, Mar. 20, 1887, Folder: Tyng Nos. 4–7, D3a 3717, Franklyn Land and Cattle Company; George Tyng to Francis de P. Foster, June 25, 1898, Folder: Statements of Account Tyng Nos. 74–77, Box 7 D. 36 3715, Franklyn Land and Cattle Company, Manuscript Collection, PPHM.

24. U.S. Bureau of the Census, *Fourteenth Census of the United States: 1920. Population Distribution of Inhabitants* (Washington, D.C.: Government Printing Office, 1921), 132, 303; U.S. Bureau of the Census, *Fifteenth Census of the United States: 1930* (Washington, D.C.: Government Printing Office, 1931), 1056, 1059, 1081; Amarillo City Commission Minutes: Jan. 3, 1925, Oct. 22, 1925, June 6, 1926.

25. Amarillo City Commission Minutes: June 10, 1926, Aug. 11, 1926, Mar. 29, 1927, Mar. 2, 1929.

26. Amarillo City Commission Minutes: Jan. 3 and Feb. 16, 1925, July 26, 1927, Aug. 23, 1927, Mar. 28, 1928. Note that declarations of water sufficiency were made roughly every ten years and continue through to present.

27. Amarillo City Commission Minutes, Apr. 28, 1925.

28. Jeanne S. Archer and Stephanie Kadel Taras, *Touching Lives: The Lasting Legacy of the Bivins Family* (Amarillo: Tell Studios, 2009), 71; Amarillo Genealogical Society, "A. S. 'Sid' Stinnett," in *Texas Panhandle Forefathers*, comp. Barbara C. Spray (Amarillo: Amarillo Genealogical Society, 1983), 161.

29. Amarillo Genealogical Society, "A. S. 'Sid' Stinnett," 161; G. Emlen Hall, *Four Leagues of Pecos: A Legal History of the Pecos Grant 1880–1933* (Albuquerque: University of New Mexico Press, 1984), 212.

30. A. E. McGregor, "Report on Reconnaissance and Preliminary Surveys of the South Canadian River for Oldham and Potter County, Texas," Folder 202, Box 5, S/N 14153, Governor Arthur Hannett Papers (NMSRCA).

31. "Preliminary Report of Commission on Drainage, Irrigation and Reclamation for Years 1925 and 1926," 8, Oklahoma State commission on Drainage, Irrigation and Reclamation, OK PR 1.1:926p (Department of State Archives, Oklahoma Department of Libraries, Oklahoma City, Oklahoma); Hutchinson County Historical Commission, *History of Hutchinson County*, 380; F. Stanley [Stanley F. L. Crocchiola], *The Stinnett, Texas, Story* (Nazareth, Tex.: Self-published, 1974); Amarillo Genealogical Society, "A. S. 'Sid' Stinnett," 161.

32. Thelma Meredith Lofgren to Terry Maxey, July 2, 1973, interview, Thelma Meredith Manuscript Collection (PPHM).

33. Jan. 7,1924, Folder1, Box 2; Sept. 9, 1926, Folder 3, Box 2; May 14 and 30, Oct. 24, 1928, Folder 5, Box 2, Ranch Manager Diaries, Alamositas Division, Matador Land and Cattle Company Collection (Southwest Collection of the Special Collections Library, Texas Tech University, Lubbock, Texas; cited hereafter as SWC-TTU).

CHAPTER 2

1. "Western Areas of State May Face Famine," and "River Rips Huge Gap in Dam: 15,000 Flee Torrent's Wrath," *Daily Oklahoman* (Oklahoma City), Oct. 16, 1923; "Preliminary Report of Commission on Drainage, Irrigation and Reclamation for Years 1925 and 1926," 8, Oklahoma State commission on Drainage, Irrigation and Reclamation, OK PR 1.1:926p (Department of State Archives, Oklahoma Department of Libraries, Oklahoma City).

2. *Tascosa Pioneer*, Sept. 15 and 22, 1886; H. C. Frankenfield, "Rivers and Floods," *Monthly Weather Review* 51 (November 1923): 604; "Western Areas of State May Face Famine," *Daily Oklahoman* (Oklahoma City),Oct. 16, 1923; Secretary to Governor Trapp to E. E. Blake, June 18, 1942, Nov. 19, 1923–Jan. 10, 1927, Series RG-8-F-2, Folder 10, Box 3a, Drainage, Irrigation and Reclamation, Governors' Papers: Martin E. Trapp (Division of State Archives, Oklahoma Department of Libraries, Oklahoma City).

3. E. E. Blake to Governor James F. Hinkle, Aug. 28, 1924, Folder 223, Box 6, S/N 14125, Governor Hinkle Papers (NMSRCA); "An Outline History of the Arkansas River," at http://www.tulsaweb.com/port/history.htm> Accessed April 24, 2012.

4. Wells A. Hutchins, *The Texas Law of Water Rights* (Austin: Texas Board of Water Engineers, 1961), 111; Donald J. Pisani, *Water, Land and Law in the West: The Limits of Public Policy, 1850–1920* (Lawrence: University Press of Kansas, 1996), 15, 7ff.

5. Pisani, *Water, Land and Law*, 7, 9, 19, 36.

6. Hutchins, *Texas Law of Water Rights*, 18, 111, 115; Amarillo City Commission Minutes, Apr. 29, 1925.

7. Hutchins, *Texas Law of Water Rights*, 111, 116, 557, 561, 567; Ira G. Clark, *Water in New Mexico: A History of Its Management and Use* (Albuquerque: University of New Mexico Press, 1987), 117, 119. The precedent the Texas court drew on was English common law, *Acton v. Blundell*, an 1843 English case revolving around the disruption of a spring by an adjacent coal mine. Looking ahead in the Canadian River's story, the Texas Supreme Court would return to this matter in 1955, *Corpus Christi v. Pleasanton* (154 TX 289), and reconfirm that if one landowner's well interfered with another landowner's well, the injured party had no recourse for damages under existing Texas laws.

8. Carl Coke Rister, *Oil! Titan of the Southwest* (Norman: University of Oklahoma Press, 1949), 280, 282, 368, 369, 370, 371; Richard R. Moore, *West Texas after the Discovery of Oil: A Modern Frontier* (Austin: Pemberton Press, 1971), x, 15–16.

9. George William Sherk, *Dividing the Waters: The Resolution of Interstate Water Conflicts in the United States* (The Hague: Kluwer Law International, 2000), 1, 3, 23; Douglas R. Littlefield, *Conflict on the Rio Grande: Water and the Law, 1879–1939* (Norman: University of Oklahoma Press, 2008), 115.

10. Canadian River Compact as printed in the Oklahoma *Senate Journal*, Mar. 23, 1927, a copy in Folder 202, Box 5, S/N 14153, Governor Hannett Papers (NMSRCA); italics added for emphasis.

11. Sherk, *Dividing the Waters*, 17; Norris Hundley, *Water and the West: The Colorado River Compact and the Politics of Water in the American West* (Berkeley: University of California Press, 1975), xv, 8, 213; John Wesley Powell, *Report on the Lands of the Arid Region of the United States with a More Detailed Account of the Lands of Utah* (2nd ed.; Washington, D.C.: U.S. Government Printing Office, 1879), 30; James E. Sherow, *Watering the Valley: Development along the High Plains Arkansas River, 1870–1950* (Lawrence: University Press of Kansas, 1990), 3, 103; 1926 Canadian River Compact, articles I and V.

12. June 10, 1929, Governor Dan Moody Veto of House Bill No. 80; *Amarillo Daily News*, Mar. 10, 1929; *Amarillo Daily News*, June 11, 1929. Governor Moody had lost a state court case over oil and gas royalties on lands in the area near Borger in 1926 and bore personal animosity toward some men in the town, which might have influenced his decision. The record does not say.

13. Andrew Hockenhull to W. A. Fogil, Sept. 29, 1933, Folder 202, S/N 13117, Andrew W. Hockenhull Papers (NMSRCA); Herbert W. Yeo, *Ninth Biennial Report State Engineer of New Mexico 1928–1930* (Santa Fe: State Engineer's Office, 1930), facing 176; Amarillo Genealogical Society, "A. S. 'Sid' Stinnett," 161.

14. Donald J. Pisani, *Water and American Government: The Reclamation Bureau, National Water Policy and the West, 1902–1935* (Berkeley: University of California Press, 2002), 237, 240, 288, 294; Donald C. Jackson, *Building the Ultimate Dam: John S. Eastwood and the Control of Water in the West* (Norman: University of Oklahoma Press, 2005), 189, 251.

15. Walter E. Parker to E. E. Blake, Feb. 17, 1928, Folder 2, Box 1 (2006.03), E. E. Blake Papers (Oklahoma State Archive, Oklahoma City); E. E. Blake to Governor Trapp, Sept. 17, 1925, Governors' Papers: Trapp; John R. Berry, *Rising Tide: The Great Mississippi Flood of 1927*

and How It Changed America (New York: Simon and Schuster, 1997), 16, 170–171, 187.

16. Amarillo Genealogical Society, "A. S. 'Sid' Stinnett," 161.

17. Pisani, *Water and Government*, 30, 52.

18. *Amarillo Daily News,* Mar. 13, 1929.

CHAPTER 3

1. *Tascosa Pioneer*, Mar. 29, 1890.

2. Hurt, *Problems of Plenty*, 61, 63; Worster, *Dust Bowl*, 94; Amity Shlaes, *The Forgotten Man: A New History of the Great Depression* (New York: Harper Collins, 2007), xiii, 7; Bonnifield, *The Dust Bowl*, 57, 58; Douglas A. Irwin, *Peddling Protectionism: Smoot-Hawley and the Great Depression* (Princeton, N.J.: Princeton University Press, 2011), 134–136, 158, 163. Irwin suggests that Smoot-Hawley did nothing to the already depressed U.S. agricultural industry.

3. Hurt, *Problems of Plenty*, 61; Worster, *Dust Bowl*, 92–94; Bonnifield, *Dust Bowl*, 37–38, 58; Rister, *Oil!*, 279, 281.

4. James C. Malin, "Dust Storms—Part I," *Kansas Historical Quarterly* 14 (May 1946), "Dust Storms—Part II," *Kansas Historical Quarterly* 14 (August 1946) , "Dust Storms—Part III," *Kansas Historical Quarterly* 14 (November 1946); *Tascosa Pioneer* Mar. 29, 1890; Apr. 24, and 25, 1912, Fol. 5, Box 1, Ranch Manager Diaries, Alamositas Division, Matador Land and Cattle Company Collection; Malin, *Winter Wheat*, 239; Priscilla H. Wilson, *A Pioneer Love Story: The Letters of Minnie Hobart* (Shawnee Mission, Kans.: TeamTech Press, 2008), 17, 23; Myrna Tryon Thomas, *The Windswept Land: A History of Moore County, Texas* (Dumas, Tex.: Self-published, 1967), 53; Bell Ranch Monthly Weather Reports Aug. 1899, June 1901, May 7, 1927, May 1930, Item Number 126, MSS 86 BC Red River Valley Corporation Collection (Center for Southwest Research, University Libraries, University of New Mexico); Bonnifield, *Dust Bowl*, 7; *Panhandle (Texas) Herald* Apr. 15, 1935; and Geoff Cunfer, *On the Great Plains* (College Station: Texas A&M University Press, 2005), 45–46.

5. Malin, *Winter Wheat*, 239, 240, 245; Thomas, *Windswept Land*, 63; Bonnifield, *Dust Bowl*, 41, 51.

6. Gary L. Nall, "Dust Bowl Days: Panhandle Farming in the 1930s," *Panhandle Plains Historical Review* 48 (1975): 43; Bonnifield, *Dust Bowl*, 57, 58;

7. Timothy Egan, *The Worst Hard Time: The Untold Story of Those Who Survived the Great American Dust Bowl* (Boston: Houghton Mifflin Co., 2006), 140, 158; author observations 1995–2014.

8. Nall, "Dust Bowl Days," 45, 51; Worster, *Dust Bowl*, 109, 110; Pamela Riney-Kehrberg, *Rooted in Dust: Surviving Drought and Depression in Southwestern Kansas* (Lawrence: University Press of Kansas, 1994), 96; A. P. Atkins, "Conservation Ranching in the Oklahoma Panhandle," *Journal of Range Management* 3 (July 1950): 167, 168; and John C. Dawson, *High Plains Yesterdays: From XIT Days through Drouth and Depression* (Austin: Eakin Press, 1985), 199.

9. Worster, *Dust Bowl*, 13; Cunfer, *On The Great Plains*, 156, 158; Paul Sears, *Deserts on the March* (3rd ed.; Norman: University of Oklahoma Press, 1959), 119–120; House Great Plains Committee, Feb. 10, 1937, H. Doc. 144, *The Future of the Great Plains*, 75th Cong., 1st Sess. 32, 34–35.

10. Worster, *Dust Bowl*, 113, 132, 133; Hurt, *Problems of Plenty*, 72–74.

11. Worster, *Dust Bowl*, 113; J. Evetts Haley, "Cow Business and Monkey Business," *Saturday Evening Post* (Dec. 8, 1934), 26, 29, 96.

12. Sean J. Flynn, "Living History: John L. McCarty and the Texas Panhandle" (Ph.D. diss., Texas Tech University, 1999), 1, 15, 28, 2.

13. Flynn, "Living History," 58; Dawson, *High Plains Yesterdays*, 215; Egan, *Worst Hard Time*, 185, 229–230.

14. Flynn, "Living History," 72, 70, 83, 89; Egan, *Worst Hard Time*, 229–30, 286; Bonnifield,

Dust Bowl, 114. It is worth noting that both John Dawson and Timothy Egan feel that John McCarty failed Dalhart by moving to Amarillo and suggest that he was a hypocrite, if not in that exact term.

15. "The Conservation Program" (editorial), *Canyon Daily News*, Dec. 30, 1937; "Panhandle Lakes Program Stimulates Business," *Southwest Business Reporter*, Sept. 1939; Flynn, "Living History," 157.

16. "Panhandle Lakes Program Stimulates Business," *Southwest Business Reporter* Sept. 1939; Flynn, "Living History," 157; "Amarillo to Be Host Today to Highway Delegates and Water Conservationists," *Amarillo Daily News*, Dec. 8, 1936; "Water Conservationists Map Organization," *Amarillo Daily News*, Dec. 9, 1936.

17. Vance Johnson, "Lakes That Dot the 'Dust Bowl' Grew from Demands of Dirt Farmers," *Amarillo Globe*, Oct. (?) 1941, Folder 49, McCarty Papers; "Panhandle Lakes Program Stimulates Business," *Southwest Business Reporter*, Sept. 1939; Minutes of Five-State Meeting of the Panhandle Water Conservation Association, May 20, 1937, p. 14, Folder 48, McCarty Papers.

18. "Forward," Fol. 202, Box 5, Hannett Papers; Arch Hurley to Andrew W. Hockenhull, Jan 30, 1934, Folder 202, S/N 13117, Hockenhull Papers; Clyde Tingley and Alverson, telephone transcript, Oct. 15, 1935, Folder 280, S/N 13102, Clyde K. Tingley Papers (NMSRCA).

19. "Contributions to Canadian Dam Fund," Folder 280, Box 8, S/N 13102, Tingley Papers.

20. Bonnifield, *Dust Bowl*, 76–77; Sears, *Deserts on the March*, 99; Richmond T. Zoch, "Rivers and Floods," *Monthly Weather Review*, 63 (May 1935): 171.

21. Vance Johnson, *Heaven's Tableland: The Dust Bowl Story* (1947; reprint, New York: De Capo Press, 1974), 198–200.

22. Tate Dalrymple et al., "Floods in the Canadian and Pecos River Basins of New Mexico, May and June 1937," USGS Water Supply Paper 842, 1939, Folder 1725, Box 69, S/N 4386, New Mexico State Engineer's Records (NMSRCA).

23. Resolution Passed by Convention Held in Little Rock February 12–13, 1937, and Report of Committee on Permanent Organization, Folder 320, McCarty Papers; Flynn, "Living History," 162, 165; Hurt, *Problems of Plenty*, 85.

24. Flynn, "Living History," 158, 159; Resolution Passed by Convention Held in Little Rock, February 12–13, 1937, Folder 320, McCarty Papers.

25. Minutes of Five-State Meeting of the Panhandle Water Conservation Association, May 20, 1937, 3, 4, 5, Folder 48, McCarty Papers.

26. Ibid., 7, 8. One wonders if the caller was J. Evetts Haley, concerned about more federal encroachment onto his and others' Panhandle property.

27. Ibid. 7, 12-a.

28. Ibid., 13, 14–15.

29. Ibid., 13, 14–15, 16, 22,; "Cannon, Clarence Andrew," *Biographical Directory of the United States Congress*, <http://bioguide.congress.gov/scripts/biodisplay.pl?index=C000117> [Accessed May 15, 2014].

30. Flynn, "Living History," 163; Bonnifield, *Dust Bowl*, 44.

31. Phillips, "FDR, Hoover and the New Rural Conservation, 1920–1932," 119, 121; Minutes of Five-State Meeting of the Panhandle Water Conservation Association, May 20, 1937, 14, Folder 48, McCarty Papers; Bonnifield, *Dust Bowl*, 6, 114, 138; Riney-Kehrberg, *Rooted in Dust*, 118.

32. Bonnifield, *Dust Bowl*, 6, 110, 112; Cunfer, *On the Great Plains*, 144, 146; "The Taylor Grazing Act," <http://www.blm.gov/wy/st/en/field_offices/Casper/range/taylor.1.html> [Accessed Jan. 25, 2014].

33 L. C. Gray, O. E. Baker, F. J. Marschner, B. O. Weitz, W. R. Chapline, Ward Shepard, and Raphael Zon, "The Utilization of Our Lands for Crops, Pasture and Forests," *USDA*

Yearbook of Agriculture 1923 (Washington D.C.: U.S. Government Printing Office, 1924), 415, 433, 505.

34. Bonnifield, *Dust Bowl,* 115–116; Riney-Kehrberg, *Rooted in Dust,* 104–105.

35. Hurt, *Problems of Plenty,* 82, 87; Bonnifield, *Dust Bowl,* 115–116; Riney-Kehrberg, *Rooted in Dust,* 104–105; Worster, *Dust Bowl,* 189–192.

36. Carl Hinton to John McCarty, Feb. 18, 1938, John McCarty to Victor Steiner (and Franck Craddock, T. H. Rixey, Roy Smith and K. C. Lee), Apr. 29, 1938, J. J. Dempsey to Franck Craddock, May 4, 1938, and Frank Craddock to John McCarty, May 9, 1938, Folder 48, McCarty Papers.

37. Minutes of Five-State Meeting of the Panhandle Water Conservation Association, May 20, 1937, 1, 2, 3, 4, 7, 10, Folder 320, McCarty Papers.

38. Minutes: Sept. 8, 1937 Meeting of the Panhandle Water Conservation Association, 13, Folder 320, McCarty Papers.

39. Ibid., 16, 17, 19, 25, 26, 30, Folder 320, McCarty Papers.

40. John L. McCarty to Olin Hinkle, Apr. 21, 1939, H. H. Finnell to John McCarty, Apr. 24, 1939, and H. H. Bennett "Land Security for the Plains," June 1938, 1, 6, Folder 48, McCarty Papers; Price and Rathjen, *Golden Spread,* 121; Flynn, "Living History," 169, 171.

41. Ross D. Rogers to Arkansas River Basin Committee and Arch Hurley, Dec. 1, 1933 in Folder 202, S/N 13117, Hockenhull Papers; *Canyon Daily News* (editorial), Dec. 30, 1937, Folder 320, McCarty Papers; Carl Hinton to John L. McCarty, Feb. 18, 1938, Folder 48, McCarty Papers.

42. Lewis Nordyke, "Conservation of Water Spreads," *Amarillo Daily News,* n.d., Folder 320, McCarty Papers; Albert Cooper to John McCarty, July 29, 1939, and C. H. Walker to John McCarty, Feb. 12, 1938, Folder 48, McCarty Papers.

43. Worster, *Dust Bowl,* 211, 219; Johnson, *Heaven's Tableland,* 223, 227; Hurt, *Problems of Plenty,* 95. R. Douglas Hurt points out another long-term change—farmers viewed assistance as a right.

44. Amarillo Genealogical Society, *Texas Panhandle Forefathers,* comp. Barbara C. Spray (Amarillo: Amarillo Genealogical Society, 1983), 161; Lofgren to Maxey, July 2, 1973, interview.

45. ."Amarillo, TX: Yearly Precipitation Totals 1892–2008," <http://www.srh.noaa.gov/ama/?n=yearly_precip> [Accessed May 15, 2014]; Green, *Land of the Underground Rain,* 125–127, 131; Bonnifield, *Dust Bowl,* 98; Bartlett, "Discovery," 53, 54; Hutchinson County Historical Commission, *History of Hutchinson County,* 286.

46. Apr. 29, May 1, May 8, May 28, June 27, Sept. 26, Oct. 15 and 21, Nov. 25, Dec. 10, 1941, Folder 3, Box 3, Ranch Manager Diaries, Alamositas Division, Matador Land and Cattle Company Records.

47. Eric C. Stene, "The Canadian River Project," <http://www.usbr.gov/projects//Image Server?imgName=Doc_1303158200779.pdf> [Accessed May 15, 2014]; Bascom N. Timmons, "Canadian River Dam at Sanford, Amarillo Urged," *Amarillo Daily News,* Aug. 6, 1941; "Dam Backers Seek Action in Capitol," *Amarillo Globe,* June 21, 1941.

48. "Menace of Onrushing Flood Water Mounting at Roswell," *Amarillo Daily News,* Sept. 4, 1941; "New Mexico Flood Peril Abates; Canadian Dam Drive Renewed," *Amarillo Daily News,* Sept. 24, 1941; L. Gifford Kessler II, "Channel Sequences and Braided Stream Development in the South Canadian River, Hutcheson, Roberts and Hemphill Counties, Texas" (Ph.D. diss., University of New Mexico, 1972), 67.

CHAPTER 4

1. U.S. Army Air Corps *Wings over America: Lubbock Army Air Field* (Baton Rouge, La.: Army and Navy Publishing, 1943), 30; "Amarillo Committeemen Mum on City's Defense Chances," *Amarillo Globe,* Aug. 29, 1941, and "Bombing Range Near Channing May Be

Secured," [source unidentified, believed to be *Amarillo Times*], Sept. 1941, Mayor Joe Jenkins Scrapbook Aug. 29, 1941–Dec. 31, 1941 (Special Collections, Central Branch Library, Amarillo Public Library, henceforth APL-C).

2. U.S. Army Air Corps, *Wings over America: Amarillo Army Air Field* (Baton Rouge, La.: Army and Navy Publishing, 1943) 55; Jan. 31, 1941, Mar. 3, 1941, May 5, 1941, Amarillo City Commission Minutes, Feb. 17, 1942, Amarillo City Commission Minutes; "Giant Bomber Base Recommended Here," *Amarillo Daily News*, Sept. 10, 1941; "City to Launch Sewer Project 'Within a Year,'" *Amarillo Globe*, Nov. 26, 1941; "About Pantex," <http://www.pantex.com/about/index.htm> [Accessed May 15, 2014].

3. "Submit $750,000 City Bond Issue," *Amarillo Globe*, Apr. 3, 1942, "City Water Plan May be Periled," *Amarillo Daily News*, May 20, 1942, Mayor Joe Jenkins Scrapbook May 14, 1942–Aug. 24, 1942 (APL-C).

4. "Water Pipe by Carload Rolling into Amarillo," *Amarillo Globe*, May 27, 1942, Jenkins Scrapbook May 14, 1942-Aug. 24, 1942.

5. May 1, 1941, Oct. 12–14, 1941, and Dec. 21, 1941, Folder 3, Box 3, Division Manager Diaries, Matador Ranch Collection (SWC-TTU).

6. Hutchinson County Historical Society, *A History of Hutchinson County* (Dallas: Taylor Publishing Co., 1980), 252, 286; Lofgren to Maxey, July 2, 1973, interview.

7. Feb. 17, 1942, Mar. 3, 1942, Apr. 2, 1941, June 26, 1942, Jan. 9, 1943, Amarillo City Commission Minutes; "Mile of Pipe Laid in a Day Here," *Amarillo Globe*, July 12, 1942; "Water May Be Rationed Here," *Amarillo Globe*, July 29, 1942.

8. "Overflowing Sewer Line Arouses Wrath of Area," *Amarillo Globe*, Aug. 9, 1946; Sept. 19, 1944, Amarillo City Commission Minutes.

9. Apr. 26, 1940, Jan. 9, 1941, Mar. 27, 1941, Sept. 10, 1942, July 13, 1944, Lubbock City Commission Minutes, vol. 6.

10. Green, *Land of the Underground Rain*, 148–149, 159, 162; J. W. Lang, *Water Resources of the Lubbock District, Texas* (Austin: Texas State Board of Water Engineers, 1945; reprint, 1953), 1, 2.

11. E. R. Leggat, *Geology and Ground Water Resources of Lynn County, Texas* (Austin: Texas Board of Water Engineers, 1952), 10, 21; Frank P. Hill and Pat Hill Jacobs, *Grassroots Upside Down: A History of Lynn County, Texas* (Austin: Nortex Press, 1986), 232, 244, 235, 241; Dawson County Historical Commission, *Dawson County History* (Lubbock: Craftsman Printers and Taylor Publishing Co., 1981), 54.

12. Hill and Jacobs, *Grassroots Upside Down*, 241; McCray and Cole, *Oil Well Drilling Technology*, 76, 48, 79; Samuel D. Myres, *The Permian Basin: Petroleum Empire of the Southwest*, vol. 2: *Era of Advancement* (El Paso: Permian Press, 1977), xii, 28, 29, 322, 363; Terry County Historical Society, *Terry County, Texas* (Clanton, Ala.: Heritage Publishing Consultants, 2002), 17; S. C. Burnitt and R. L. Crouch, *Investigation of Ground Water Contamination, P.H.D., Hackberry and Storie Oil Fields, Garza County, Texas* (Austin: Texas Water Commission, 1964), 9, 17, 24.

13. Mar. 7, 1944, Amarillo City Commission Minutes, vol. 12; W. A. Warren to John McCarty, Apr. 27, 1942, Folder 320, McCarty Papers.

14. "City May Use Conchas Water," clipping from unknown paper, June 30, 1942, Jenkins Scrapbooks May 14, 1942-Aug. 24, 1942.

15. "Overflowing Sewer Line Arouses Wrath of Area," *Amarillo Globe*, Aug. 9, 1946; "Amarillo Gets Sand with her Tap Water These Days, but It's Good, Clean Sand," *Amarillo Globe*, Aug. 9, 1946.

16. "Water Crisis," *Amarillo Globe*, June 18, 1946; "Water Supply Is Adequate," *Amarillo Globe*, June 23, 1946; "These Times," *Amarillo Times*, Apr. 8, 1947. The *Times* was the first tabloid in the state, and its editorial views were more skeptical about certain city policies and plans than were those of the *Globe* and *Daily News*.

17. Lewis Nordyke, "Water: Well Expert Outlines Own Plan for Amarillo Supply," *Amarillo Globe*, Aug. 4, 5, 1946; "Water: The Globe Supports Bond Issue, with Reservations," *Ama-*

rillo Globe, Sept. 16, 1946; "Mayor Appoints Citizen Group," *Amarillo Times*, Sept. 16, 1946; "City Will Be Host to Secretary Krug," *Amarillo Globe*, Oct. 4, 1946; "Rainwater Overloaded Sewer Lines," *Amarillo Globe*, Oct. 7, 1946; "Plains Highways Gradually Drain Following Flood," *Amarillo Daily-News*, Oct. 8, 1946. The *Daily News* was the morning edition of the *Globe* and owned by the same people.

18. Nordyke, "Water"; "These Times," *Amarillo Times*, Apr. 8, 1947; "City to Go Southwest for Additional Water," *Amarillo Globe*, June 25, 1947.

19. Green, *Land of the Underground Rain*, 78–79; D. L. McDonald, "The People Are Talking about Amarillo Water," *Amarillo Globe*, Apr. 18, 1947; D. L. McDonald, "So The People May Know about Water," *Amarillo Times*, May 11 and 18, 1947; T. E. Johnson, "These Times," *Amarillo Times*, May 25, 1947; "Water Supply Key to the Future Growth of Amarillo," D. L. McDonald reprinting Frank Langston from *Dallas Times-Herald*, in *Amarillo Globe*, June 15, 1947.

20. Cal Brumley, "City's Dream of Plenty of Water Soon to Be Realized," *Amarillo Globe*, Oct. 16, 1947; "City Bid for Water Challenged," *Amarillo Globe*, Jan. 9, 1948; "Randall Farmers Delay Protests of Water Site," *Amarillo Times*, Jan. 21, 1948; Cal Brumley, "Embattled Randall Farmers May Ask Amarillo to Pay for Water," *Amarillo Globe*, Jan. 28, 1948; "City Ready to Condemn Land," *Amarillo Times*, Jan. 29, 1948; "Second Suit Blocks City Water Plans," *Amarillo Globe*, Feb. 2, 1948.

21. "Three-State Water Board to Develop Canadian River Resources Urged," *Amarillo Globe-News*, Oct. 14, 1947; "Worley's Aid Sought on Canadian Dam Project," *Amarillo Globe*, Apr. 3, 1948.

22. Amarillo Genealogical Society, *Texas Panhandle Forefathers*, 161; Stene, "Canadian River Project"; Bascom N. Timmons, "Canadian River Dam at Sanford, Amarillo Urged," *Amarillo Daily News*, Aug. 6, 1941; "Dam Backers Seek Action in Capitol," *Amarillo Globe*, June 21, 1941.

23. "Mayor Gains Support in Fight for Dam," *Amarillo Globe*, Apr. 19, 1948; "Hagy Outlines Steps for Obtaining Canadian Dam," *Amarillo Times*, Apr. 19, 1948.

24. Hutchinson County, *Hutchinson County History*, 253, 286; George Finger to Phil D. Phillips July 18, 1965, interview, Folder: Ms/Int: 1977–31/12, Manuscript Collection (PPHM); George Tyng to F. de P. F, June 25, 1898, Folder: Tyng Statement of Accounts, Box 7, Francklyn Land and Cattle Company Records, D.36 3715, Manuscript Collection, PPHM; Winifred Vigness, "Municipal Government in Lubbock," in *A History of Lubbock*, ed. Lawrence L. Graves (Lubbock: West Texas Museum Association, 1962), 383, 384; C. C. Reeves Jr. and Judy A. Reeves, *The Ogallala Aquifer of the Southern High Plains*, vol. 1: *Geology* (Lubbock: Estacado Books, 1996), 156, 232.

25. "Hereford Mobilizes to Fight for Canadian Dam at Tascosa," *Amarillo Times*, May 10, 1949; H. B. Virgil Crawford to George Mahon, Apr. 27, 1949, Folder 7, Box 486, George Mahon Papers (SWC-TTU).

26. "South Plains Joins Amarillo in Canadian River Dam Project," *Amarillo Globe*, May 19, 1948; "Spur Dam Project," *Amarillo Globe*, June 30, 1948; "Plans for Dam in High Gear," *Amarillo Times*, June 30, 1948.

27. Lofgren to Maxey, July 2, 1973, interview.

28. "Plan Backers Push for Canadian Dam," *Amarillo Daily News*, July 1, 1948; "Lubbock in Dam Plea," *Amarillo Globe*, Aug. 24, 1948.

29. Minutes: Meeting of Tri-State Governors' Committee–South Canadian River Amarillo Texas Aug. 3, 1948, Folder 241, McCarty Papers; "Lubbock and Amarillo Begin Dam Survey," *Amarillo Daily News*, Oct. 5, 1948; Burt C. Blanton, "Survey Portraying Need for Dam on Canadian River Prepared by Amarillo Chamber of Commerce and City of Lubbock," Dec. 1, 1948, Folder 8, Box 486, Mahon Papers; Stene, "Canadian River Project."

30. "Lubbock and Amarillo Begin Dam Survey," *Amarillo Daily News*, Oct. 5, 1948; D. W. Britain to John McCarty, July 2, 1948, Folder 241, McCarty Papers.

31. Amarillo City Commission Minutes, Oct. 19, 1948, electronic document no. 948957 (also available in original bound volumes or through request to city secretary) (City Secretary's Office, City of Amarillo, Amarillo, Texas); "Dam Prospects Bright," *Amarillo Globe*, Nov. 19, 1948; "Canadian River Bill under Federal Study," *Amarillo Globe*, Feb. 17, 1949; "Boost Dam at Meeting," *Amarillo Daily News*, Mar. 3, 1949; "Water Needs from Canadian Outlined," *Amarillo Globe*, Mar. 8, 1949.

32. "Water Needs from Canadian Outlined," *Amarillo Globe*, Mar. 8, 1949; Finger to Phillips, July 18, 1965, interview.

33. Reconnaissance Report: Canadian River Project Texas, Project Planning Report No. 5-12.22-0, June 1949, Folder 243, McCarty Papers.

34. Clarence Whiteside to George Mahon, Sept. 17, 1949, H. J. Keifer, "River Project Benefits Seen," *Lubbock Avalanche*, June 18, 1949, J. T. Hutchinson, M.D. to George Mahon, July 2, 1949, Folder 7, Box 486, Mahon Papers; Amarillo City Commission Minutes, Oct. 19, 1948, electronic doc no. 948957; Amarillo City Commission Minutes, July 15, 1949, electronic doc no. 949090; Thomas Thompson, "Future Drab with No Dam," *Amarillo Daily News*, July 7, 1949.

35. "Canadian River Damsite at Sanford," *Amarillo Daily News*, Apr. 21, 1949; B. Byron Price, "Haley, James Evetts, Sr.," *The Handbook of Texas Online*, <http://www.tshaonline.org/handbook/online/articles/fhahj> [Accessed May 15, 2014]; Haley, "Cow Business and Monkey Business;"; "Worley, Francis Eugene," *Biographical Directory of the United States Congress*, <http://bioguide.congress.gov/scripts/biodisplay.pl?index=W000744> [Accessed May 15, 2014]. The Fain family also owned property near the Amarillo dam site, and the newspaper may have been confused.

36. John McCarty to Joe Wolters, Nov. 27, 1948, Folder 241, McCarty Papers W. V. McCoy, "Hereford Mobilizes for a Fight," *Amarillo Times*, May 10, 1949; "Tascosa Dam-Site Is Urged," [source uncertain, believed to be from *Amarillo Globe*], May 11, 1949, Folder 241, McCarty Papers; Elster M. Haile to George Mahon, July 5, 1949, Levelland Chamber of Commerce to George Mahon May 27, 1949, telegram, and H. B. Crawford to George Mahon Apr. 27, 1949, Folder 7, Box 486, Mahon Papers.

37. House Committee on Public Lands, July 8, 1949, Hearings before a Subcommittee on Irrigation and Reclamation of the Committee on Public Lands on H. R. 2733, 81st Cong., 1st Sess. (Serial 18).

38. T. E. Johnson, "These Times," *Amarillo Times*, June 22, 1949; "Find Prospects for Dam Rosy," *Amarillo Globe*, July 3, 1949; "Dam Backers Seek Action in Capitol," *Amarillo Globe*, June 21, 1949; Thomas Thompson, "Congressional Subcommittee Acts Swiftly, Oklahoma Ayes Canadian Dam," *Amarillo Globe*, July 8, 1949.

39. Stene, "Canadian River Project."

40. "Dam Bill Reaches Senate," *Amarillo Globe-News*, Aug. 15, 1949; "New Mexico Gives Green Light on Dam," *Amarillo Daily News*, Sept. 7, 1949; Elizabeth Carpenter, "Anderson Delays Hearings on Dam," *Amarillo Times*, Aug. 16, 1949. The author has been unable to determine thus far if it was Sen. Forrest C. Donnell or Sen. James P. Kern.

41. Elizabeth Carpenter, "Anderson Delays Hearings on Dam," *Amarillo Times*, Aug. 16, 1949; "Denies Blocking Dam," *Amarillo Globe*, Aug. 17, 1949.

42. "Anderson's Stand Not Over-Alarming," *Lubbock Avalanche*, Aug. 20, 1949; "Canadian Dam Backers in Talks with Anderson," *Amarillo Times*, Sept. 4, 1949; "New Mexico Gives Green Light on Dam," *Amarillo Daily News*, Sept. 7, 1949; L. R. Hagy to L. B. Johnson, Sept. 9, 1949, Folder 7, Box 486, Mahon Papers.

43. L. A. Wilkie to Henry Teubel, Aug. 18, 1949, Folder 7, Box 486, Mahon Papers, "Gift Horse Could Be a Poor Bargain," *Fort Worth Star Telegram*, Sept. 13, 1949.

44. A. A. Meredith to Dennis Chavez, Apr. 29, 1950, A. A. Meredith to Arthur V. Watkins, n.d., and "Senator Blocks Canadian Project," *Plainview Evening Herald*, Aug. 9, 1950.

Notes to Pages 83–89

45. George Mahon to Charlie Guy, Dec. 15, 1950, Folder 6, Box 486, Mahon Papers; "The Canadian Dam: A New Slant," *Lubbock Avalanche*, Jan. 19, 1951, and A. A. Meredith to George Mahon, Jan. 9, 1951, Folder 7, Box 486, Mahon Papers.

CHAPTER 5

1. A. A. Meredith to George Mahon, Feb. 14, 1953, Folder 4, Box 486, Mahon Papers, and *Floyd County Hesperian*, Jan. 15, 1953.

2. Minutes of Tri-State Governors' Committee–South Canadian River, Aug. 3, 1948, Folder 241, McCarty Papers; Clarence Burch to Michael W. Strauss, Aug. 18, 1949, Folder 242, McCarty Papers; Sherk, *Dividing the Waters*, 10.

3. Sherow, *Watering the Valley*, 116, 117.

4. Canadian River Compact, article I; Clark, *Water In New Mexico*, 541; *States of Oklahoma and Texas, Plaintiffs, v. State of New Mexico*, No. 109, 501 US 221 (1991), footnote 4; cited hereafter as *Oklahoma et al. v. New Mexico*.

5. Canadian River Compact, articles III, IV, V, VII and IX; Clark, *Water in New Mexico*, 541; Sherk, *Dividing the Waters*, 16.

6. Texas Board of Water Engineers, *21st Annual Report Board of Water Engineers for the State of Texas Covering Biennium September 1, 1952 to August 23, 1954* (Austin: Texas Board of Water Engineers, 1954), xii, xx, cited hereafter as TBWE, *21st Report*; Texas Board of Water Engineers, *22nd Annual Report Board of Water Engineers for the State of Texas Covering Biennium September 1, 1954 to August 31, 1956* (Austin: Texas Board of Water Engineers, 1956), x, cited hereafter as TBWE, *22nd Report*; Stene, "Canadian River Project." Although not exactly blackmail and not illegal, such pre-claims to project water bedeviled attempts to reclaim the arid West, going back to the Powell Survey era.

7. "The Canadian Dam: Here Is a New Slant," *Lubbock Avalanche*, Jan. 2, 1951, and "Panhandle Delegation Ired at Stand by Connally, Johnson on Dam Funds," *Fort Worth Star Telegram*.

8. "Canadian River Group to Study Ways to Speed Big Project," *Lubbock Evening Journal*, Aug. 6, 1951; "Panhandle Delegation Ired at Stand by Connally, Johnson on Dam Funds," *Fort Worth Star Telegram*, Aug. 25, 1951.

9. TBWE, *21st Report*, x, xii; Hutchins, *Texas Law of Water Rights*, 12, 111, 112; Marc Roy Rasor, "Canadian River Water District Approved by Board of Engineers," *Lubbock Avalanche*, Mar. 7, 1953.

10. Canadian River Municipal Water Authority Enabling Act, Art. 8280–154 (Acts 1953, 53rd Leg. Ch. 243), cited hereafter as CRMWA Enabling Act.

11. Morris G. Cobb to Debra Wood Oct. 3, 1990, Folder Ms/Int: Cobb, Morris, 1991–157/19, Manuscript Collection, (PPHM).

12. Stene, "Canadian River Project"; TBEW, *22nd Report*, x; CRMWA Enabling Act, RVSD 2009, 1, 5.

13. CRMWA Enabling Act, 1, 3, 5.

14. Enabling Legislation as Amended: Acts 1987, 70th Leg., ch. 251, par. 2, emerg. eff. May 28, 1987.

15. CRMWA Enabling Act, 7.

16. Clarence K. Whiteside, "Facts about the Canadian River Project," Nov. 1953, Canadian River Project Reference File (SWC-TTU). Proration in this context meant restrictions and rationing.

17. Abbe et al., *Lubbock and the South Plains*, 39, 41, 57, 69; Green, *Land of the Underground Rain*, 109–110.

18. Lubbock City Commission Minutes, Apr. 26, 1956, and Mar. 28, 1957; "A Brief History of Lubbock Water Supplies," and Muleshoe Chamber of Commerce to Baldridge Bakery of Lubbock, Apr. 21, 1956, File 3.50415 et al. (Lubbock Municipal Water Authority, Lubbock,

Texas); "Bailey Opens Water Fight," *Lubbock Evening Journal*, Mar. 23, 1956, and W. H. Graham Jr. "Lubbock Eyes Plains Sandhills for Future Water Supply Needs," *State Line Tribune* (Parmer County), Mar. 22, 1956.

19. "Amarillo, TX: Yearly Precipitation Totals 1892–2008," <http://www.srh.noaa.gov/ama/?n=yearly_preci> [accessed May 14, 2014]; "Lubbock Precipitation 1911–Present," <http://www.srh.noaa.gov/lub/?n=climate-klbb-pcpn> [Accessed May 15, 2014]; "Dwindling Water Supplies Threaten Disaster for Nation," *Lubbock Evening Journal*, Sept. 26, 1955; Martin Melosi, *The Sanitary City: Urban Infrastructure in America from Colonial Times to Present* (Baltimore: Johns Hopkins University Press, 2000), 226; Myres, *Permian Basin*, 216, 252, 322, 364; Rister, *Oil!*, 282, 369, 370. The author's mother had younger relatives from Central Texas who were four and five years old before they saw rain.

20. Howard E. Robbins to T. E. Johnson, n.d., Canadian River Project Reference File.

21. "Haley Defends Stand on Dam," *Sun* (Lubbock) Jan. 25, 1953.

22. Ibid.

23. A. A. Meredith to George Mahon, Feb. 14, 1953, Tracey Campbell to George Mahon, Jan. 17, 1953, and "Haley Defends Stand on Dam," *Sun* (Lubbock) Jan. 25, 1953.

24. The author has found no evidence either way thus far and is making a supposition. Bureau of Reclamation, *Land Acquisition Process for Federal Reservoirs* (pamphlet), Jan. 1961, Amarillo City Commission Records, electronic doc no. 935799.

25. CRMWA Minutes: Oct. 11, 1954, 1, 2, Folder 4, Box 486, Mahon Papers.

26. George Finger to Phil D. Phillips, Feb. 18, 1965, 7–9, Folder Ms/Int: 1977–31/12; Reconnaissance Report Canadian River Project, Project Planning Report No. 5–12.22–0 (June 1949), 20, 24, McCarty Papers; CRMWA Minutes: Oct. 11, 1954, 1, 2, Folder 4, Box 486, Mahon Papers; "22 Million Savings Seen if Cities Build Canadian Project," *Lubbock Avalanche*, Dec. 18, 1954.

27. "22 Million Savings Seen if Cities Build Canadian Project," *Lubbock Avalanche*, Dec. 18, 1954, and "Dam Not Dead, Yet," *Lubbock Avalanche-Journal*, Jan. 12, 1955; "Notices Posted on Canadian Water Project," *Plainview Sunday Herald*, May 29,1955; Amarillo City Commission Minutes, Mar. 13, 1955.

28. Amarillo City Commission Minutes, Feb. 2 and July 28, 1955; "22 Million Savings Seen if Cities Build Canadian Project," *Lubbock Avalanche*, Dec. 18, 1954,; "Notices Posted on Canadian Water Project," *Plainview Sunday Herald*, May 29, 1955; A. A. Meredith to George Mahon, Feb. 12, 1955, Folder 2, Box 486, Mahon Papers; "Amarillo Officials to Explain Possible 'Pullout' on Dam," *Borger Daily Herald*, July 22, 1955; "Amarillo Asks Authority to Offer Alternative Plan," *Borger Daily Herald*, Aug. 17, 1955.

29. "Borger Only 'Yankee' Left," *Borger Daily Herald*, Aug. 3, 1955; "Public Vote on Dam Promised," Ed Deswysen, *Borger Daily Herald*, July 19, 1955; Hill and Jacobs, *Grassroots Upside Down*, 232, 241.

30. "Cities OK Canadian Dam Bond Issue by 11–1 Margin," *Borger News-Herald*, Nov. 9, 1955; "Plainview Rejects Canadian Dam," *Lubbock Avalanche*, Apr. 27, 1956.

31. "Amarillo, Pampa Citizens Get Last Chance to Vote on Dam," *Borger News-Herald*, Aug. 9, 1955; "Eight Cities OK Dam Contract; Plainview Falls Out," *Borger News Herald*, Mar. 14, 1956; Finger to Phillips, Feb. 18, 1965, 6, 7.

32. "Decision on Canadian River Contract Due Tuesday," *Lamesa Daily Reporter*, Sept. 1, 1961.

33. "Texas Legislature Approves CRMWA Water Rate Change," [clipping from unspecified source, believed to be *Lubbock Avalanche*], Apr. 26, 1957, and A. A. Meredith to George Mahon, Aug. 24, 1957, Folder 26, Box 485, Mahon Papers; C. W. Ratliff, "Canadian Dam Directors Back New 4-Zone Water Price Set Up," *Lubbock Avalanche*, Dec. 16, 1958; Finger to Phillips, Feb. 18, 1965, interview; "Amarillo Officials Slow Dam Speed," *Borger News-Herald*, Feb. 12, 1958.

34. Stene, "Canadian River Project"; Canadian River Project: Letter from the Secretary of the Interior, Aug. 4, 1950, iv, 13–14, Folder 22, Box 485, Mahon Papers; "Decision on Canadian River Contract Due Tuesday," *Lamesa Daily Reporter*, Sept. 1, 1961.

CHAPTER 6

1. Claire E. Swan, *Scottish Cowboys and the Dundee Investors* (Dundee, Scotland: Aberty History Society, 2004), 11, 12; Robert H. (R.H.) Fulton to Jeff Townsend, June 14, 1972, interview, Fulton, R.H., Oral History Collection (SWC-TTU).

2. Cal Farley and Elvon L. Howe, *Ten Thousand Sons: The Story of Cal Farley's Boys' Ranch* (Canaan, N.H.: Phoenix Publishing, 1987), 5, 126–228; Marcia Caldwell to Jewell F. Pendleton, Nov. 17–20 1975, and Jan. 6–9, 1976, interview (transcript), 2–3, 5, 19 footnotes 2 and 3, Ms/Int File 1976–34/8, Manuscript Collection (PPHM).

3. Paul H. Carlson, *Empire Builder in the Texas Panhandle: William Henry Bush* (College Station: Texas A&M University Press, 1996), 34, 107, 121; Archer and Taras, *Touching*, 92, 154, 201.

4. Farley and Howe, *Ten Thousand Sons*, 166; Caldwell to Pendleton, Nov. 17–20 1975, and Jan. 6–9, 1976, interview, 1, 2.

5. "Yearly Rainfall Totals," <http://www.srh.noaa.gov/ama/?n=yearly_precip> [Accessed May 14, 2014]; the years 1970 and 2011 would break this record.

6. "Yearly Rainfall Totals"; "Cold and Warm Episodes by Season," <http://www.cpc.ncep. noaa.gov/products/analysis_monitoring/ensostuff/ensoyears.shtml> [Accessed May 15, 2014]; Ethredge, "Role of the North American Monsoon, 73–74. This pattern is referred to the El Niño–Southern Oscillation, or ENSO.

7. Lasley and Holt (comps.), *Dust Storms and Half Dug-Outs*, 48, 109, 165, 315; Nellie Witt Spikes, *As a Farm Woman Thinks: Life and Land on the Texas High Plains, 1890–1960*, ed. Geoff Cunfer, Plainsword by Sandra Scofield (Lubbock: Texas Tech University Press, 2010), 116, 121.

8. Malin, "Dust Storms—Part One," 130, 132; Malin, "Dust Storms—Part Two," 267, 271, 280; author observations, 1990–2011.

9. Bickers, "Three Cultures, Four Hooves and One River," 178–180. See Appendix B for methodology and charts.

10. Carlson, *Amarillo*, 165; Spikes, *Farm Woman*, 108; Melosi, *Sanitary City*, 226. The ongoing Cold War may have been another consideration, in that the federal government may not have wanted to play up any drought-related hardships lest they turn into propaganda one way or another.

11. Ken E. Rogers, *The Magnificent Mesquite* (Austin: University of Texas Press, 2000), 18, 20, 48; C. F. Fisher, "The Mesquite Problem in the Southwest" *Journal of Range Management* 3, (January 1950): 60, 62.

12. Fisher, "Mesquite Problem," 61, 62; Rogers, *Magnificent Mesquite*, 23, 55; Henry A. Wright and Arthur W. Bailey, *Fire Ecology of the United States and Southern Canada* (New York: John Wiley and Sons, 1982), 28, 90.

13. J. W. Dollahite and W. V. Anthony, *Nutrition in Cattle on an Unbalanced Diet of Mesquite Beans*, Texas A&M Ag Experiment Station Progress Report #1931 (College Station: Texas A&M University, 1957), 1, 3; Texas Agricultural Experiment Station, *Mesquite: Growth, Development, Management, Economics, Uses* (College Station: Texas A&M University, Texas Agricultural Experiment Station, 1973), 7. The high protein in mesquite beans eventually overloads bovine intestinal microbes, interfering with digestion of grasses and grains if the cattle eat too much mesquite for too long.

14. Fisher, "Mesquite Problem," 60, 62; John Cypher, *Bob Kleberg and the King Ranch: A Worldwide Sea of Grass* (Austin: University of Texas Press, 1996), 75.

15. Jerome S. Horton, "The Development and Perpetuation of the Permanent Tamarisk Type in the Phreatophyte Zone of the Southwest," 124, 125, <http://www.rmrs.nau.edu/awa/

ripthreatbib/horton_devperptamarisk.pdf> [Accessed May 15, 2014]; T. H. Moser, "What Has Been Done about Salt Cedar at Caballo Reservoir," *Reclamation Era* 46 (May 1960): 37, 38; Clark, *Water in New Mexico*, 391–392; Samuel Sikes and Jackie Smith, "A Vegetational Study of the Canadian River Breaks," in *The Canadian River Breaks: A Natural Area Survey*, part 7 of 7 (Austin: University of Texas Press, 1975), 48.

16. Moser, "Salt Cedar," 39, 40; Clark, *Water in New Mexico*, 132.

17. S. A. Schumm and R. W. Lichty, *Channel Widening and Flood-Plain Construction Along Cimarron River in Southwestern Kansas*, 79, 82, 86; VanLooy and Martin, "Channel and Vegetation Change," 727, 736–737.

CHAPTER 7

1. C. W. Ratliff, "Barbecue Crowd Gets Canadian Dam Report," *Lubbock Avalanche-Journal*, May 22, 1961; George Mahon to H. S. Hilburn, July 10, 1962, and "Canadian River Project Groundbreaking Ceremony at Vista Pint, Canadian River Dam Site," pamphlet, Folder 21, Box 485, Mahon Papers.

2. Walter Rogers to A. A. Meredith, June 24, 1961, Folder 22, Box 485, Mahon Papers "Fate of Canadian River Project in Doubt Despite Presidential Appropriation Recommendation," *Lubbock Sunday Sun*, January 29, 1961.

3. Wilbur Rogers to George Mahon, Sept., 1961, Folder 22, Box 485, Mahon Papers; Clark, *Water in New Mexico*, 325–326; David Remley, *Bell Ranch: Cattle Ranching in the Southwest, 1824–1947* (rev. ed.; Las Cruces, N.M.: Yucca Tree Press, 2000), 289, 295.

4. Jackson, *Building the Ultimate Dam*, 14, 19, 20, 189; Ken Weaver, *Dam Foundation Grouting* (New York: American Society of Civil Engineers, 1991), 1, 15, 16; Rupert C. S. Walters, *Dam Geology* (London: Butterworths and Co., 1962), 4, 9, 11; Stene, "Canadian River Project."

5. Walters, *Geology*, 15; Stene, "Canadian River Project"; "Canadian River Project," <http://www.usbr.gov/projects//ImageServer?imgName=Doc_130315820079.pdf> [Accessed May 15, 2014].

6. V. O. Grantham, "Innovations in Aqueduct Construction on the Canadian River Project," *Reclamation Age* 50 (August 1964): 53; W. J. Schaefer to George Mahon, May 24, 1963, Folder 20, Box 485, Mahon Papers; "Canadian River Project Faces Congress Hurdle," *Amarillo Times*, June 17, 1949.

7. Ned Curran, "Canadian River Funds Controversy Revealed," *Lubbock Avalanche-Journal*, Sept. 12, 1963; Stene, "Canadian River Project"; Bascom N. Timmons, "Canadian River Dam at Sanford, Amarillo Urged," *Amarillo Daily News*, Aug. 6, 1941; "Dam Backers Seek Action in Capitol," *Amarillo Globe*, June 21, 1941.

8. Flynn, "Living history," 235; "Meredith Given Award by Interior Department," *Plainview Daily Herald*, Mar. 11, 1963; Lofgren to Maxey, July 2, 1973, interview.

9. Gifford Pinchot, "The Fight for Conservation," in *Conservation in the Progressive Era: Classic Texts*, ed. David Stradling (Seattle: University of Washington Press, 2004), 21; John Wesley Powell, *Report on the Arid Region of the United States with a More Detailed Account of the Lands of Utah*, ed. Wallace Stegner (1879; reprint, Cambridge, Mass.: Harvard University Press, 1962), 21, 43, 52; Samuel P. Hays, *Conservation and the Gospel of Efficiency: The Progressive Conservation Movement, 1890–1920* (Cambridge, Mass.: Harvard University Press, 1959), 5, 266.

10. A. A. Meredith to John F. Kennedy, July 11, 1962, in Folder 21, Box 485, Mahon Papers.

11. Samuel P. Hays and Barbara D. Hays, *Beauty, Health and Permanence: Environmental Politics in the United States 1955–1985* (New York: Cambridge University Press, 1987), xi, 3, 13, 22, 65, 139; Adam Rome, *The Bulldozer in the Countryside: Suburban Sprawl and the Rise of American Environmentalism* (New York: Cambridge University Press, 2001), 6, 7, 146, 127–128; Nash, *Wilderness and the American Mind*, 192, 225.

12. Hays and Hays, *Beauty, Health and Permanence*, 1, 3, 35, 41; Karl Boyd Brooks, *Before*

Earth Day: The Origins of American Environmental Law, 1945–1970 (Lawrence: University Press of Kansas, 2009), 64–65, 94; "Cuyahoga River Fire," <http://www.ohiohistorycentral.org/entry.php?rec=1642> [Accessed May 15, 2014]. The river had also burned in 1952, drawing nothing but local attention.

13. Donald Worster, *Rivers of Empire: Water, Aridity and the Growth of the American West* (New York: Oxford University Press, 1985), 149, 335; Charles E. Wilkinson, *Crossing the Next Meridian: Land, Water and the Future of the West* (Washington, D.C.: Island Press, 1992), 234, 285–286; Norris Hundley Jr., *Water and the West: The Colorado River and the Politics of Water in the American West* (2nd ed.; Berkeley: University of California Press, 2009), 347–348; Hal K. Rothman, *Saving the Planet: The American Response to the Environment in the Twentieth Century,* (Chicago: Ivan R. Dee, 2000), 102, 105.

14. Worster, *Rivers of Empire*, 106-7, 273-74; Wilkinson, *Crossing the Next Meridian*, 275–276; the definitive account of the Grand Canyon preservation struggle is Byron E. Pearson, *Still the Wild River Runs: Congress, the Sierra Club and the Fight to Save Grand Canyon* (Tucson: University of Arizona Press, 2002); for an extended discussion of the personalities involved, see John McPhee, *Encounters with the Archdruid* (New York: Farrar, Straus and Giroux, 1971).

15. Arthur Carhart, "Turn Off That Faucet!" *Atlantic* 185 (February 1950): 39, 41, 42; Wallace Stegner, "Myths of the Western Dam," *Saturday Review,* Oct. 23, 1965, 29, 31. Carhart's article was reprinted in the April 1950 edition of *Reader's Digest.*

16. William Bowen, "Water Shortage Is a Frame of Mind," *Fortune* 71 (April 1965): 144, 145, 146, 190, 195, 198.

17. "History of CRMWA"<http://crmwa.com/history-of-crmwa/> [Accessed May 15, 2014].

18. New Mexico State Engineer's Office, *Preliminary Report on the Geology of the Ute Dam Site, Quay County, New Mexico* (Santa Fe: State Engineer's Office, Technical Division, 1961), Fol. 1001, Box 29, S/N 4346, New Mexico State Engineer's Records (NMSRCA), 1, 34; Frank E. Wozniak, *Across the Caprock: A Cultural Resources Survey on the Llano Estacado and the Canadian River Valley of Eastern New Mexico for the Bravo CO2 Pipeline* (Albuquerque: Office of Contract Archaeology, University of New Mexico, 1985), 12–13; "Ute Lake State Park," <http://geoinfo.nmt.edu/tour/state/ute_lake/home.html> [Accessed May 17, 2014].

19. Charles Robert Kelly, "The Canadian River Municipal Water Authority Project in West Texas: A Geographic Analysis" (Ph.D. dissertation, University of Oklahoma, 1971), 152, 154, 158–161.

20. Kelly, *Canadian River*, 139; Grantham, "Innovations in Aqueduct Construction," 53, 55; Eugene B. Williams to George Mahon May 16, 1966, Fol. 17, Box 485, Mahon Papers.

21. Minutes of CRMWA Board of Directors Meeting, Aug. 13, 1973; Minutes of CRMWA Board of Directors Meeting, Jan. 8, 1975, 14; Minutes of CRMWA Board of Directors Meeting, Oct. 14, 1968, 11.

22. Minutes of CRMWA Board of Directors Meeting, Oct. 14, 1968, 11; "Tastes Nasty," W. D. B. to Editor, *Amarillo Globe-Times*, July 18, 1968; "Lake Water Shutoff Would Increase Tab," *Amarillo Globe-Times*, July 3, 1968; "Amarillo Water Tops Rival Cities in Taste Test," *Amarillo Globe-Times*, July 17, 1968; "Bait Dealer: 'Lake Water Ruined Me,'" *Amarillo Globe-Times*, July 31, 1968. The letters to the editor section of the *Globe-Times* featured at least one water letter every other day during the period May–August, most of them related to the flavor.

23. Gordon R. Forsyth, "Reclamation's 11-City Water Pipe," *Reclamation Era* 52 (February 1966): 8; Program: Dedication of Sanford Dam November 1, 1966, Folder 17, and W. Rogers to George Mahon, Dec. 9, 1966, Folder 14, all in Box 485, Mahon Papers.

24. Gordon R. Forsyth, "Reclamation's 11-City Water Pipe," *Reclamation Era* 52 (February 1966): 8, and Program: Dedication of Sanford Dam November 1, 1966, Folder 17, Box 485, Mahon Papers.

CHAPTER 8

1. "Order AAFB Phased Out," *Amarillo Daily News*, Nov. 20, 1964.

2. "Amarillo Air Force Base Scrapbook—1956–1969," Archives (APL-C).

3. Carlson, *Amarillo*, 179, 180; "Order AAFB Phased Out," *Amarillo Daily News*, Nov. 20, 1964; "Our History," <http://www.aaf-hf.org/about/history.htm> [Accessed Apr .22, 2012]; "Amarillo Hospital District," <http://www.amarillohospitaldistrict.org> [Accessed Apr. 22, 2012]; "These Times," *Amarillo Times*, Apr. 8, 1947.

4. CRMWA Minutes: Mar. 18, 1968, 9; Oct. 14, 1968, 4.

5. CRMWA Minutes: Oct. 14, 1968, 4, 5; July 14, 1969, 2, 4, 5.

6. CRMWA Minutes: Apr. 13, 1970, 4, 5, 6, 9, 15, 18.

7. CRMWA Minutes: July 13, 1970, 24–25, Appendix A. Of thirty pages of minutes, nineteen cover the budgeting and billing debates.

8. CRMWA Minutes: Oct. 12, 1970, 31, 32; "Summary of Events Affecting Canadian River Project Facilities Related to the Lubbock Tornado of May 11, 1970," Folder "Correspondence, 1960–1984 and Undated," Box 1, I. F. Lea Papers (SWC-TTU); "Lubbock Tornado," <http://www. Lubbockonline.com/lubbocktornado.shtml> [Accessed May 15, 2014].

9. CRMWA Minutes: Oct. 14, 1968, 10; Aug. 14, 1969, 11; Melosi, *Sanitary City*, 163–164.

10. Presley, "Salt Deposition Systems," 24, 25, 28 fig; T. C. Gustavson, Robert C. Finley, and Robert W. Baumgardner Jr., "Preliminary Rates of Slope Retreat and Salt Dissolution along the Eastern Caprock Escarpment of the Southern High Plains and in the Canadian River Valley," in *Geology and Geohydrology of the Palo Duro Basin, Texas Panhandle: A Report on the Progress of Nuclear Waste Isolation Feasibility Studies* 1979, ed. T. C. Gustavson et al., Geology Circular 80-7 (Austin: University of Texas Press for the Bureau of Economic Geology, 1980), 76, 88; Joe H. Smith, "The Plains Plazas: A Brief Pause in the Span of Time," *Southwest Heritage* 3, No. 2 (1969).

11. CRMWA Minutes: Jan. 12, 1970, 8, 9. Lake turnover means the upper and lower layers of the lake reverse position. Most common in deep, narrow lakes in the fall and in the spring.

12. CRMWA Minutes: Jan. 11, 1971, 6; Oct. 8, 1971, 9; May 15, 1972, 3.

13. Zane Spiegel, "Natural Discharge of Saline Water from Permian Rocks in the Basins of the Canadian, Red, Brazos and Colorado Rivers in New Mexico, Texas and Oklahoma," General Report, Aug. 1966, 1, 2, 7, Folder Misc. Docs 1363, New Mexico State Engineer's Records (NMSRCA); Bureau of Reclamation, "Lake Meredith Salinity Study Canadian River Texas-New Mexico Appraisal Level Investigation," Oct. 1979, 3, 9, 18, 20, Canadian River Project, Box 65, 8N5 11595083, RG 115 (National Archives and Records Administration, Denver, Colorado; hereafter cited NARA-Denver).

14. CRMWA Minutes: Apr. 23, 1973, 5–7; Oct. 15, 1973, 12.

15. CRMWA Minutes: May 14, 1973, 2; July 9, 1973, 2; Oct. 15, 1973, 12.

16. CRMWA Minutes: Apr. 11, 1978, 2; Bureau of Reclamation, "Lake Meredith Salinity Study," 18, 20.

17. Green, *Land of the Underground Rain*, 170, 191.

18. Green, *Land of the Underground Rain*, 192–193; John Opie, *Ogallala: Water for a Dry Land* (2nd ed.; Lincoln: University of Nebraska Press, 2000), 208.

19. "Zybach, Frank," in *The Grasslands of the United States: An Environmental History*, ed. James E. Sherow (Santa Barbara, Calif.: ABC-Clio, 2007), 338–339; Opie, *Ogallala*, 142–144.

20. Gunnar Brune, *Springs of Texas* (2nd ed. College Station: Texas A&M University Press, 2002), 246, 247, 347, 350.

21. Ibid., 369, 349. Brune provides liters per second flows, which were converted to liters per year, then gallons per year. CRMWA Minutes: July 9, 1986, 4.

22. Brune, *Springs of Texas*, 12; Paul L. Younger, *Groundwater in the Environment: An Introduction* (Malden, Mass.: Blackwell Publishing, 2007), 113, 114, 115, 116.

23. U.S. Department of the Interior, Fish and Wildlife Service, "A Detailed Report on

the Fish and Wildlife Resources Sanford Reservoir, Canadian River Project," May 1954, 7, 9, 10, 18, Box 128, FRC224 504, CRP-565.00–54–05 CI, Bureau of Reclamation Files (NARA-Denver); "Fishing Lake Meredith," at "Lake Meredith," <http://www.tpwd.state.tx.us/fishboat/fish/recreational/lakes/meredith/> [Accessed May 15, 2014]; Wes Phillips, "Fish of Lake Meredith," <http://www.panhandlenation.com/geographica/fish/fish.html> [Accessed Apr. 24, 2012]; CRMWA Minutes: Oct. 14, 1968, 4.

24. Bureau of Reclamation, "Project Histories: Canadian River Project Texas Vol. IX Calendar Years 1970–1976," 1, 31–33, Box 266, 115–87–0008 88N-115–92–130, and Bureau of Reclamation, "Canadian River Project History Vol. X Calendar Years 1977–1980," 20, Box 16, 8NS 115–93–213, RG 115 (NARA-Denver).

25. CRMWA Minutes: Oct. 14, 1968, 4.

26. CRMWA Minutes: Oct. 12, 1970, 21; Apr. 23, 1973, 8; July 9, 1973, 5; Oct. 15, 1973, 12; Jan 14, 1974, 10; July 12, 1974, 10; Oct. 14, 1974, 16; Apr. 9, 1975, 9; July 9, 1975, 16–17; July 14, 1976, 14; Oct. 13, 1976, 9; Bureau of Reclamation, "Canadian River Project Texas Vol. IX Calendar Years 1970–1976," 23.

27. CRMWA Minutes: Jan. 17, 1969, 3, 4; Oct. 14, 1974, 17; Bureau of Reclamation, "Canadian River Project History Vol. X Calendar Years 1977–1980," 19.

28. Hurt, *Problems of Plenty*, 132–133; John Erickson, *The Modern Cowboy* (Lincoln: University of Nebraska Press, 1981) 204–207; J.C. McNeill III, *The McNeills' SR Ranch: 100 Years in Blanco Canyon* (College Station: Texas A&M University Press, 1988), 186.

29. CRMWA Minutes: Apr. 9, 1975, 3; Apr. 13, 1977, 13, 14; Worster, *Rivers of Empire*, 325–326; Mark Reisner, *Cadillac Desert: The American West and Its Disappearing Water* (rev. and updated ed.; New York: Penguin Books, 1993), 316–323.

30. "Digest of Federal Resource Laws of Interest to the U.S. Fish and Wildlife Service: Federal Water Pollution Control Act (Clean Water Act)," <http://www.fws.gov/laws/lawsdigest/fwatrpo.html> [Accessed June 23, 2014]; CRMWA minutes, Jan. 11, 1974, Jan. 11, 1978; Ricky George, "Water Pollution Expert Retiring" *Amarillo Globe-News,* June 1, 2001, <http://amarillo.com/stories/2001/06/01/new_expert.shtml> [Accessed June 23, 2014].

31. CRMWA Minutes, July 14, 1969; Jan. 12, 1970; Apr. 23, 1973; May 14, 1973; Oct. 15, 1973; Bureau of Reclamation, "Canadian River Project Texas Vol. IX Calendar Years 1970–1976," 18, 28, 26, 25; Reisner, *Cadillac Desert*, 402–403.

32. Stene, "Canadian River Project"; CRMWA Minutes: July 7, 1972; Jan. 22, 1973; Apr. 6, 1973; Jan. 8, 1975; July 13, 1977.

33. CRMWA Minutes: Apr. 9, 1980; Bureau of Reclamation, "Canadian River Project Texas Vol. IX Calendar Years 1970–1976," 7, 13.

34. David P. Hill, "Mammals of the Lake Meredith Recreation Area and Alibates National Monument" (Master's thesis, West Texas State University, 1984), 5, 28, 31, 40, 42, 48, 49, 66, 69; Bureau of Reclamation, "Canadian River Project Texas Vol. IX Calendar Years 1970–1976," 7, 13.

35. Bill Cox, "Court Reverses Judgment against Water Authority," *Amarillo Daily News* Nov. 19, 1974, clipping, Bureau of Reclamation "Canadian River Project Texas Vol. IX Calendar Years 1970–1976"; CRMWA Minutes: Jan. 11, 1971, 7; Jan. 14, 1974, 7, 8.

36. CRMWA Minutes: Jan. 8, 1975, 14, 15, 16; Vaughn Hendrie, "Water Ruling Saves Lubbock $6 Million," *Lubbock Avalanche-Journal*, Nov. 20, 1974, 1, 12, clipping, Bureau of Reclamation, "Canadian River Project Texas Vol. IX Calendar Years 1970–1976," RG 115.

37. CRMWA Minutes: Jan. 8, 1975; Aug. 3, 1973.

38. CRMWA Minutes: July 8, 1968, 2; Oct. 13, 1976, 13; Apr. 11, 1978, 2–3.

39. CRMWA Minutes: July 9, 1975, 6; Oct. 8, 1975, 10; Apr. 14, 1976, 5; July 14, 1976, 3; July 13, 1977, 6–19, 22.

40. CRMWA Minutes: Apr. 13, 1977.

CHAPTER 9

1. CRMWA Minutes: Jan. 9, Apr. 8, Oct. 8, 1980; Jan. 14, Sept. 8, 1981; Jan. 13, 1982.

2. CRMWA Reports: Oct. 14, 1981; Jan. 13, July 14, Oct. 12, 1982; Jan. 11, Apr. 12, Oct. 12, 1983.

3. CRMWA Reports: Jan. 10, Apr. 12, July 11, 1984; Jan. 7, Oct. 7, 1986; Jan. 13, July 7,1987; July 12, 1988; July 11, 1990; "Desalinization of Brackish Water and Seawater," <http://www.prominent.com/desktopdefault.aspx?tabid-190/570_read-2273/> [Accessed May 14, 2014].

4. Amarillo, TX: Yearly Precipitation Totals 1892–2008" <http://www.srh.noaa.gov/ama/?n=yearly_precip >[Accessed May 14, 2014]."Cold and Warm Episodes by Season" <http://www.cpc.ncep.noaa.gov/products/analysis_monitoring/ensostuff/ensoyears.shtml> [Accessed May 15, 2014]; CRMWA Records, Jan. 7, 1985; Jan. 10, 1990.

5. "Ute Lake State Park" <http://geoinfo.nmt.edu/tour/state/ute_lake/home.html> [Accessed May 17, 2014]; Bureau of Reclamation, "Final Environmental Statement, Eastern New Mexico Water Supply Project, New Mexico" (FES 76–47, 1976) A2, New Mexico State Engineer's Records (NMSRCA); Canadian River Compact, article IV (a) (b).

6. "Reservoir Terminology," <http://midgewater.twdb.state.tx.us/swrweb/swr/hydro/Hydro_Definitions.html> [Accessed Apr. 24, 2012].

7. CRMWA Reports: Apr. 10, July 11, 1984; Philip B. Mutz, "Reservoir Storage Development in the Canadian River Basin and Related Litigation," in *Water Issues of Eastern New Mexico: Get Your Water Kicks on Route 66* (Las Cruces: New Mexico Water Resources Research Institute, 1997), 14; U.S. Army Corps of Engineers, "Report on Sedimentation Conchas Reservoir Canadian River Basin Resources of October 1963," Albuquerque District of the Army Corps of Engineers, Dec. 1966, 33, New Mexico State Engineer's Records (NMSRCA); *Oklahoma et al. v. New Mexico*, 15, 16, 18.

8. Mutz, "Reservoir Storage," 14–15.

9. Hutchins, *Texas Law of Water Rights*, 17, 25–26, 128.

10. Garnett P. Williams and M. Gordon Wolman, *Downstream Effects of Dams on Alluvial Rivers*, USGS Professional Paper 1286 (Washington, D.C.: U.S. Government Printing Office, 1984), 31, 33, 35, 47, 51; Charles W. Martin and William C. Johnson, "Historical Channel Narrowing and Riparian Vegetation Expansion in the Medicine Lodge River Basin, Kansas, 1871–1983," *Annals of the Association of American Geographers* 77, no. 3 (1987): 442, 443, 447; Schumm and Lichty, *Channel Widening and Flood-plain Construction Along Cimarron River in Southwestern Kansas*, 73, 79, 86–87.

11. Rosgen, *Applied River Morphology*, 2; Martin and Johnson, "Historical Channel Narrowing," 442–443.

12. Decision in *E. H. Brainard, II, Carolyn Rogers, Nancy Briscoe, Boone Pickens, Bea Pickens, Morrison Cattle Company, J. A. Whittenburg, III, Frances W. Klein, Jack F. Turner, Diane E. Bowes, and J. A. Whittenburg, IV, et al., Petitioners v. The State of Texas and the General Land Office of the State of Texas, Respondents*, Texas Supreme Court No. 98–0578, cited hereafter as *Brainard et al.*, <http://www.supreme.courts.state.tx.us/historical/1999/oct/980578.pdf> [Accessed May 14, 2014]; Hutchins, *Texas Law of Water Rights*, 29, 30; Rosgen, *Applied River Morphology*, 2–4; Luna B. Leopold, *A View of the River* (Cambridge, Mass.: Harvard University Press, 1994), 9, 128, 141.

13. Charles C. Mann and Mark L. Plummer, *Noah's Choice: The Future of Endangered Species* (New York: Alfred A. Knopf, 1995), 150, 160, 161, 175; Charles E. Gilliland, *Endangered Species Act: Impact on Texas Rural Land Values* (College Station: Texas Real Estate Center, 1995), 1, 4–5.

14. William L. Graf, *Wilderness Preservation and the Sagebrush Revolution* (Savage, Md.: Rowman and Littlefield Publishers, 1990), xii, 203, 231, 232; Robert H. Nelson, *Public Lands and Private Rights: The Failure of Scientific Management* (Savage, Md.: Rowman and Littlefield, 1995), 171.

15. Graf, *Wilderness Preservation*, 225, 228, 250.

16. Nash, *Wilderness and the American Mind*, 336; Graf, *Wilderness Preservation*; 232, 257; "Sierra Club Timeline–History," <http://www.sierraclub.org/history/timeline.aspx> [Accessed May 17, 2014].

17. Linda Campbell, *Endangered and Threatened Animals of Texas: Their Life History and Management* (Austin: Texas Parks and Wildlife Resource Protection Division, 1995), iv; Jody Sheets, "A Brief Commentary on the Abuse of Sovereign Immunity by the State of Texas along Texas Rivers," <http://www.texas-wildlife.org/index.php?option=com_content&view=article&id=44:a-brief-commentary-on-the-abuse-of-sovereign-immunity-by-the-state-of-texas-along-texas-rivers&catid=16:water&Itemid=> [Accessed Jan. 28, 2014]. Jody Sheets did note the rising price of oil in his article.

18. CRMWA Records: Jan. 14, 1981, 3; Oct. 11, 1989; Jan. 10, 1990; Apr. 11, 1990; Jan. 8, 1991; Sheets, "Brief Commentary"; "Canadian River Boundary Dispute Enters Round Two," <http://www.livestockweekly.com/papers/96/12/19/19river.asp> [Accessed Apr. 24, 2012].

19. CRMWA Reports: July 11, 1984, 2; July 12, 1988, 2. Italics in original.

20. CRMWA Reports: Apr. 10, 1984; July 12, 1988; July 11, 1990; Apr. 10, 1991.

21. Omer C. Stewart, *Forgotten Fires: Native Americans and the Transient Wilderness,* ed. Henry T. Lewis and M. Kat Anderson (Norman: University of Oklahoma Press, 2002), 114, 144; Paul H. Carlson, "Indian Agriculture, Changing Subsistence Patterns and the Environment of the Southern Great Plains," *Journal of Agricultural History* 66,(Spring 1992): 59; CRMWA Reports: July 12, 1974; July 9, 1975; Author conversation with USFS personnel, Lake Meredith National Recreation Area, March 2011.

22. CRMWA Records: Apr. 7, 1987; July 7, 1987.

23. CRMWA Records: July 1991; Apr. 7, 1987.

24. Reisner, *Cadillac Desert*, 444, 446–448, 488–489; Amarillo City Commission Minutes, Feb. 25, 1969, electronic doc no. 873278. The records of Water, Inc. were donated to Texas Tech University, where they wait for someone to write the full history of the group.

CHAPTER 10

1. For a description of the duties of a special master and full-time river master, see "Delaware River Master," <http://water.usgs.gov/osw/odrm/decree.html> [Accessed May 15, 2014]; Jerome C. Mays, "Special Master Report, Canadian River Delivered October 12, 1990," 2–4.

2. Mays, "Special Master Report," 92.

3. Ibid., 92, 93, 97.

4. "A Celebration of New Mexico's First Women Lawyers," <http://www.nmbar.org/Attorneys/BarsLegalGroups/NMWomensBar/nmwomenbarfirstfemaleattys.html> and <http://www.nmbar.org/Attorneys/BarsLegalGroups/NMWomensBar/womanbarfirst100.pdf. [Accessed May 15, 2014]; *States of Oklahoma and Texas, Plaintiffs v. State of New Mexico*, 501 US 221 (1991); Mays, "Special Master Report," 24.

5. Mays, "Special Master Report," 39; *Oklahoma et al. v. New Mexico*, para. 13, 21.

6. *Oklahoma et al. v. New Mexico*, para. 21–24. Emphasis in the original.

7. Ibid., para. 29–32, and footnotes 4, 5, 6.

8. Ibid., para. 45, 46, 50, 52.

9. Ibid., para. 36, 37.

10. Sherk, *Dividing the Waters*, 16, 17, 18, 59, 66.

11. CRMWA Quarterly Reports: Jan. 7, 1992;, Apr. 7, 1992, July 8, 1992, Jan. 12, 1993, Oct. 12, 1993, Jan. 11, 1994; Sherk, *Dividing the Waters*, 373.

12. CRMWA Reports: Oct. 9, 1990; "Amarillo TX: Yearly Precipitation Totals 1892-2008" <http://www.srh.noaa.gov/ama/?n=yearly_precip> [Accessed May 17, 2014]; Rusty Hawkins, "John C. Williams and West Texas Reclamation," *Panhandle Plains Historical Review* 77 (2004): 40–41; multiple Amarillo residents, 1995–present. The author has heard versions of this joke in many parts of western Texas.

13. John C. Williams, "Eleven Thirsty Cities Seeking Water Now and into the Future," *Panhandle Plains Historical Review* 78 (2004): 52; Williams and Wolman, *Downstream Effects of Dams on Alluvial Rivers*, 35, 47, 51; Martin and Johnson, "Historical Channel Narrowing," 442, 443, 447; Schumm and Lichty, *Channel Widening and Flood-plain Construction along Cimarron River in Southwestern Kansas*, 79, 86.

14. "Water Use in Santa Fe 2009." Accessed via "Reports and Studies related to water management," <http://www.santafenm.gov/how_much_water_do_we_use_reports_and_studies> [Accessed May 15, 2014]; "Water Restrictions in New Mexico Help Shore Up Dwindling Supplies," <http://www.uswaternews.com/archives/arcconserv/6newmex.html> [Accessed Apr. 24, 2012]; and Cassie McDuff, "Inland Cities Could Learn from Santa Fe., N.M.'s Water Restrictions," *Riverside Press-Enterprise*, Oct. 22, 2007, <http://www.pe.com/columns/cassiemacduff/stories/PE_News_Local_B_bcass23.3f2bdc8.html>[Accessed Apr. 24, 2012].

15. Hawkins, "John C. Williams," 40; Williams, "Eleven Thirsty Cities," 54–55; "Lake Alan Henry Water Supply Project," <http://www.team-psc.com/sɪwa_lah.html> [Accessed May 17, 2014].

16. Hawkins, "John C. Williams," 35, 36.

17. Reisner, *Cadillac Desert*, 244–248.

18. "Population History of Selected Cities, 1850–2000," <http://www.texasalmanac.com/topics/population> [Accessed May 17, 2014]; Mark Odintz, Roberts County," *The Handbook of Texas Online*, <http://www.tshaonline.org/handbook/online/articles/hcr08> [Accessed June 13, 2011]. John R. Erickson, *Through Time and the Valley*, gives a good history of the Breaks in this part of the Panhandle.

19. Williams, "Eleven Thirsty Cities," 53; Hawkins "John C. Williams," 40; History of CRMWA," <http://crmwa.com/history-of-crmwa/> [Accessed May 15, 2014]; "Testimony of John C. Williams, General Manager, Canadian River Municipal Water Authority before the Water and Power Resources Subcommittee House Resources Committee July 29, 1997," <http://naturalresources.house.gov/UploadedFiles/John_C._Williams_testimony_7.29.97.pdf> [Accessed May 13, 2014].

20. CRMWA Reports: Oct. 11, 1989; Hawkins, "John C. Williams" 41; "Testimony of John C. Williams," 2.

21. "Testimony of John C. Williams," 2–3; "History of CRMWA." Accessed May 17, 2014.

22. "Testimony of John C. Williams," 2–3.

23. Hawkins, "John C. Williams," 42–43.

24. Hawkins, "John C. Williams," 43; United States House of Representatives Committee on Resources, Oct. 6, 1998, H. Report 105-785, Conveyance of Various Reclamation Projects and Facilities—Report Together with Dissenting Views, 105th Cong., 2nd sess., 9, 10, 29.

25. Hawkins, "John C. Williams," 43, 44; Williams, "Eleven Thirsty Cities," 52, 53.

26. Author observation, Mar. 2011. This description is not intended to in any way encourage trespassing in the riverbed.

27. Sheets, "Brief Commentary"; *Brainard et al.*

28. *Oklahoma v. Texas* 260 U.S. 606 (1923) at 632. http://supreme.justia.com/cases/federal/us/260/606/ [Accessed May 17, 2014].

29. *Oklahoma v. Texas* 260- 606 (1923) at 632; *Brainard et al.*, 4, 13, 19;

30. Texas GLO website maps; *Brainard et al.*, 6; Michael V. Powell "Riparian Boundaries in Texas," <http://www.lockelord.com/~/media/Files/NewsandEvents/News/2006/03/Riparian%20Boundaries%20in%20Texas/Files/Riparian/FileAttachment/Riparian%20Boundaries%20-%20Powell.pdf> [Accessed May 15, 2014].

31. *Brainard et al.*, 5, 6.

32. *Brainard et al.*, 6–8.

33. *Brainard et al.*, 8–9; Arthur Stiles, "The Gradient Boundary—The Line between Texas and Oklahoma along the Red River," *Texas Law Review* 30 (1952): 305, 310 at "The Gradi-

ent Boundary," <http://www.tpwd.state.tx.us/publications/nonpwdpubs/water_issues/rivers/navigation/riddell/gradientboundary.phtml.> [Accessed Jan. 28, 2014].

34. "Oral Argument–2/10/99 98–0578 *E. H. Brainard II v. State of Texas*," <http://www.supreme.courts.state.tx.us/oralarguments/transcripts/98-0578.pdf.> [Accessed May 13, 2014]. Note that this transcription records Michael V. Powell's name as "Powers" and Mary A. Keeney as "Keeny."

35. "Oral Argument," 8–12. Emphasis added.

36. "Oral Arguments," 12–15.

37. *Brainard et al.*, 30, 31, 33, 35, 39, 41.

38. "Canadian River Boundary Dispute Enters Round Two"; David Bowser, "Protestors Demand Rehearing on Canadian River Decision," <http://www.livestockweekly.com/papers/99/11/04/whlcanadian.asp> [Accessed Apr. 24, 2012].

Conclusion

1. Pinchot, "The Fight for Conservation," 22; *Ladies Home Journal*, "Conservation," in Stradling, *Conservation in the Progressive Era*, 33–34.

2. Robert V. O'Neill, "Is It Time to Bury the Ecosystem Concept? (With Full Military Honors, of Course!)," *Ecology* 82, no. 12 (2001): 3277, 3279, 3282; Frederick E. Clements, *Plant Succession: An Analysis of the Development of Vegetation*, Carnegie Institute of Washington Publication 242 (Washington: Gibson Brothers Press, 1916), 3–5.

3. And not simply the Canadian watershed: residents of the Oklahoma Panhandle in the Dry Cimarron watershed express dissatisfaction with decisions made in Oklahoma City and for the same reasons.

4. Wilkinson, *Crossing the Next Meridian*, 17.

Bibliography

Primary Sources:

Abert, William. *Expedition to the Southwest: An 1845 Reconnaissance of Colorado, New Mexico, Texas and Oklahoma.* Lincoln: University of Nebraska Press, 1999.

"Aeronautics: Amarillo Air Force Base." Clippings File. Amarillo Public Library Central Branch, Amarillo, Texas.

Amarillo Air Force Base Scrapbook. Amarillo Public Library Central Branch, Amarillo, Texas.

Amarillo, Texas, City Commission Minutes, 1925–1965.

Amarillo Daily News. 1909–2001.

Amarillo Globe. 1924–1951.

Amarillo Globe-News. 2001-present. The combined Sunday edition of the *Daily News* and *Globe Times* was the *Amarillo Globe-News,* prior to the 2001 full merger of the two papers.

Amarillo Globe-Times. 1957–2001.

Amarillo Times. 1937–1951.

Atkins, A. P. "Conservation Ranching in the Oklahoma Panhandle." *Journal of Range Management* 3, No. 3 (July 1950): 167–171.

Bell Ranch Monthly Weather Reports, 1899–1955. Red River Valley Corporation Collection, Center for Southwest Research, University of New Mexico, Albuquerque, New Mexico.

Borger (Texas) News-Herald. 1925–1970.

Bowen, William. "Water Shortage Is a Frame of Mind." *Fortune,* April 1965.

Burnett, Lucien, to Woody Coffee. Manuscripts/Interview Collection: Burnett, Lucien. Panhandle-Plains Historical Museum Archive, Canyon, Texas (henceforth PPHM).

Burnitt, S. C., and R. L. Crouch. "Investigation of Ground Water Contamination P.H.D., Hackberry and Storie Oil Fields, Garza County, Texas." Austin: Texas Water Commission, 1964.

Butler, Ira. "The Oil and Gas Industry and Water Conservation." Report given at the 16th Annual Institute of Oil and Gas Law and Taxation, Dallas, Tex., February 11, 1965.

Caldwell, Marcia, to Jewell F. Pendleton. Manuscript/Interview Collection, file 1976-34/8. PPHM.

Canadian River Compact of 1926.

Canadian River Municipal Water Authority Minutes, 1969–1991. CRMWA Main Offices, Sanford, Texas.

Canadian River Project Folder. Southwest Collection/Special Collections Library, Texas Tech University, Lubbock, Texas.

Canyon Daily News 1937.

Carhart, Arthur. "Turn off that Faucet!" *Atlantic*, February 1950. 39–42.

Clyde K. Tingley Papers. New Mexico State Record Center and Archive, Santa Fe, New Mexico.

Daily Oklahoman (Oklahoma City). 1923.

Dawson, John C. *High Plains Yesterdays: From XIT Days through Drouth and Depression.* Austin: Eakin Press, 1985.

E. E. Blake Papers. Oklahoma State Historical Society, Oklahoma City, Oklahoma.

E. H. Brainard II, Carolyn Rogers, Nancy Briscoe, Boone Pickens, Bea Pickens, Morrison Cattle Company, J. A. Whittenburg III, Frances W. Klein, Jack F. Turner, Diane E. Bowes AND J. A. Whittenburg IV, et al., Petitioners v. The State of Texas and the General Land Office of the State of Texas, Respondents. Texas Supreme Court No. 98-0578. <http://www.supreme.courts.state.tx.us/historical/1999/oct/980578.pdf> [Accessed May 14, 2014].

Ellis, Ida Mae. "So That History May Not Repeat." *Reclamation Era*, Feb. 1965. 12–13.

Farley, Cal, and Elvon L. Howe. *Two Thousand Sons: The Story of Cal Farley's Boys' Ranch.* Caanan, N.H: Phoenix Publishing, 1987.

Finger, George, to Phil D. Philips. Folder Ms/Int: 1977-31/12. Manuscript Collection, PPHM.

Follett, C. R. *Records of Water Level Measurements in Cochran, Crosby, Gaines, Hockley, Lynn and Terry Counties.* Austin: Texas Board of Water Engineers, 1954.

Forsyth, Gordon R. "Reclamation's 11-City Water Pipe." *Reclamation Era*, February 1966. 9–12.

Franklyn Land and Cattle Company Papers. PPHM.

Fulton, R. L. Oral History. Southwest Collection/Special Collections Library, Texas Tech University, Lubbock, Texas.

Governor Toney Anaya Papers. New Mexico State Research Center and Archive, Santa Fe, New Mexico.

George Mahon Papers. Southwest Collection-Special Collections, Texas Tech University, Lubbock, Texas.

Gray, L. C., O. E. Baker, F. J. Marschner, B. O. Weitz, W. R. Chapline, Ward Shepard, and Raphael Zon. "The Utilization of Our Lands for Crops, Pasture and Forests." In *USDA Yearbook of Agriculture 1923*. Washington D.C.: U.S. Government Printing Office, 1924. 415–516.

Grentham, V. O. "Innovations in Aqueduct Construction on the Canadian River Project." *Reclamation Era,* Aug. 1964. 53–55.

Grozier, Richard U, Harold W. Albert, James F. Blakey, and Charles H. Hembree. Report 22. *Water Delivery and Low Flow Studies Pecos River, Texas Quantity and Quality 1964 and 1965,* <http://www.twdb.state.tx.us/publications/reports/numbered_reports/doc/R22/r22_pecosriver1966.pdf> [Accessed June 21, 2014]

———. Report 76. *Water Delivery Study Pecos River, Texas Quantity and Quality, 1967.* Austin: Texas Water Development Board, May 1968, <https://www.twdb.state.tx.us/publications/reports/numbered_reports/doc/R76/R76.pdf> [Accessed June 21, 2014]

Hagy, Lawrence. Mayoral Scrapbooks. Central Branch Library, Amarillo Public Libraries, Amarillo, Texas.

Haley, J. Evetts. "Cow Business and Monkey Business." *Saturday Evening Post,* Dec. 8, 1934, 26–29, 94–95.

———. "Haley Defends Stand on Dam." *Lubbock Sun,* Jan. 25, 1953.

Hannett, Arthur. Papers. New Mexico State Record Center and Archive, Santa Fe, New Mexico.

High Plains Study Council. *A Summary of Results of the Ogallala Aquifer Regional Study with Recommendations to the Secretary of Commerce and Congress.* Dec. 1982.

Hinkle, James. Papers. New Mexico State Record Center and Archive, Santa Fe, New Mexico.

Hockenhull, Andrew W. Papers. New Mexico State Record Center and Archive, Santa Fe, New Mexico.

Holbrook, Winfield. Papers. Manuscript/Interview Collection, PPHM.

Ingle Forrest Lea Papers. Southwestern Historical Collection-Special Collections, Texas Tech University, Lubbock, Texas.

J. Evetts Haley Papers. J. Evetts and Nita Stewart Haley Memorial Library and History Center, Midland, Texas.

Jenkins, Joe. Mayoral Scrapbooks. Central Branch Library, Amarillo Public Libraries, Amarillo, Texas.

John L. McCarty Papers. Bush-Fitzsimon Special Collection, Central Library, Amarillo Public Libraries, Amarillo, Texas.

Lang, J. W. *Water Resources of the Lubbock District, Texas.* Austin: Texas Board of Water Engineers, 1945; reprint, 1953.

Lasley, Bob, and Sallie Holt (eds.). *Dust Storms and Half Dug-Outs: Tales from the Good Old Days in the Texas Panhandle.* Hickory, N.C.: Hometown Memories Publishing Co., 2009. (Second printing entitled *Dust Storms and Half Dug-Outs: A Living History of the Texas Panhandle.* All other information unchanged.)

Lofgren, Thelma Meredith. "Interview of Thelma Meredith Lofgren as Given to Terry Maxey on July 2, 1973." Manuscript/Interview Collection. PPHM.

Lubbock Avalanche-Journal. 1940–1990.

Lubbock, Texas, City Commission Minutes, City of Lubbock Offices, City Municipal Building, Lubbock, Texas. 1925–1960.

Martin E. Trapp Papers. Division of State Archives, Oklahoma Department of Libraries, Oklahoma City, Oklahoma.

Matador Ranch Collection, Alamositas Division. Southwest Collection-Special Collections, Texas Tech University, Lubbock, Texas.

Mays, Jerome C. "Special Master Report, Canadian River, Delivered October 15, 1990," <http://www.supremecourt.gov/specmastrpt/ORG%20109%20101590.pdf> [Accessed June 21, 2014]

McNeill, J. C.., III. *The McNeills' SR Ranch: 100 Years in Blanco Canyon.* College Station: Texas A&M University Press, 1988.

Mead, Ellwood. *Irrigation Institution: A Discussion of the Economics and Legal Questions Created by the Growth of Irrigated Agriculture in the West.* 1903; reprint, New York: Arno Press, 1972.

Montaignes, François des. *The Plains.* Edited by Nancy Alpert Mower and Don Russell. Norman: University of Oklahoma Press, 1972.

Morris Cobb Papers. Manuscript/Interview Collection, PPHM.

Moser, T. H. "What Has Been Done about Salt Cedar at Caballo Reservoir." *Reclamation Era*, May 1960. 37–40.

Oklahoma State Commission on Drainage, Irrigation and Reclamation. Department of State Archives, Oklahoma Department of Libraries, Oklahoma City, Oklahoma.

"Oral Argument—2/10/99 98-0578 *E. H. Brainard II v. State of Texas,*" <http://www.supreme.courts.state.tx.us/oralarguments/transcripts/98-0578.pdf.> [Accessed May 13, 2014].

Panhandle Geological Society. *Oil and Gas Fields of the Texas and Oklahoma Panhandles.* [Amarillo]: Panhandle Geological Society, 1961.

Panhandle Herald. 1935.

Plainview Evening Herald. 1941–1970.

Pinchot, Gifford. "The Fight for Conservation." In *Conservation in the Progressive Era: Classic Texts,* ed. David Stradling. Seattle: University of Washington Press, 2004. 19–23.

Powell, John Wesley. *Report on the Arid Region of the United States with a More Detailed Account of the Lands of Utah.* 2nd ed. Washington, D.C.: U.S. Government Printing Office, 1879.

———. *Report on the Arid Region of the United States with a More Detailed Account of the Lands of Utah.* Edited by Wallace Stegner. 1879; reprint, Cambridge, Mass.: Harvard University Press, 1962.

Romero, José Ynocencio, to Ernest R. Archambeau. "Spanish Sheepmen on the Canadian at Old Tascosa." *Panhandle Plains Historical Quarterly* 19 (1946): 45–72.

Office of New Mexico State Engineer Files. New Mexico State Records

Center and Archive, Santa Fe, New Mexico.

Office of New Mexico State Engineer Papers. New Mexico State Engineer's Library, Santa Fe, New Mexico.

Erwin R. Smith Photograph Collection, J. Evetts and Nita Stewart Haley Memorial Library and History Center, Midland, Texas.

Southwest Business Reporter. 1939.

State of Oklahoma and State of Texas vs. State of New Mexico. 501 U.S. 221 (1991).

State of Texas, Plaintiff-Appellant v. Reuben Pankey, Jim Brown, Marcus Burks, W. F. Martin, Frank Sauble, T. L. Roach, Carl Hennegan and Dewey Gann, Defendants-Appellees. 44 F2d 236 (1971).

Stegner, Wallace. "Myths of the Western Dam." *Saturday Review,* October 23, 1965. 29–31.

Tascosa Pioneer. 1886–1887.

Taylor, Howard D., and Staff. *Water Level Data from Observation Wells in Southern High Plains of Texas, 1971–1977.* Austin: Texas Department of Water Resources, 1979.

Texas Board of Water Engineers. *Annual Reports.* 1950–1965.

Texas Water Development Board. *A Summary of the Preliminary Plan for Proposed Water Resources Development in the Canadian River Basin.* Austin: Texas Water Development Board, 1966.

U.S. Army Air Corps. *Wings over America: Amarillo Army Air Field.* Baton Rouge, La.: Army and Navy Publishing, 1943.

———. *Wings over America: Lubbock Army Air Field.* Baton Rouge, La.: Army and Navy Publishing, 1943.

U.S. Bureau of the Census. *Fourteenth Census of the United States: 1920. Population Distribution of Inhabitants.* Washington, D.C.: Government Printing Office, 1921.

———. *Fifteenth Census of the United States: 1930.* Washington, D.C.: Government Printing Office, 1931.

U.S. Department of Commerce. Census, 1910.

———. Census, 1920.

———. Census, 1930.

———. Census. Vol. 7: *Agriculture 1909–1910.* Washington, D.C.: U.S. Government Printing Office, 1913.

———. *United States Census for Agriculture 1920.*

———. *United States Census for Agriculture 1935.* Vol. 1. Washington, D.C.: U.S. Government Printing Office, 1936.

U.S. Department of Commerce. U.S. Weather Bureau. *Monthly Weather Review.* 1904–1980.

U.S. Department of the Interior. U.S. Bureau of Reclamation. "Eastern New Mexico Water Supply Project Feasibility Report." May 1972.

———. "Eastern New Mexico Water Supply Special Environmental Report." March 1993.

———. "Final Environmental Statement: Eastern New Mexico Water Supply

Project, New Mexico." FES 76-47, 1976.

———. "Lake Meredith Salinity Study Canadian River Texas–New Mexico Appraisal Level Investigation." October 1979.

———. *Reconnaissance Report: Canadian River Project Texas*. Project Planning Report No. 5-12.22-0, June 1949.

———. U.S. Bureau of Reclamation, Southwest Division. "Canadian River Project." File RG 115, U.S. Bureau of Reclamation Files, Southwest Division, National Archives and Record Center, Denver, Colorado.

U.S. House of Representatives. October 6, 1998. H. Report 105-785, *Conveyance of Various Reclamation Projects and Facilities—Report Together with Dissenting Views*. 105th Cong., 2nd sess.

———. Committee on Public Lands. July 8, 1949. *Hearings before a Subcommittee on Irrigation and Reclamation of the Committee on Public Lands on H. R. 2733*. 81st Cong., 1st Sess. (Serial 18).

———. Great Plains Study Committee. Feb. 10, 1937. H. Doc. 144, *The Future of the Great Plains*. 75th Congress.

Vaudrey, Walter C. *Floods of May 1955 in Colorado and New Mexico*. U.S.G.S. Water Supply Paper 1455-a. Washington, D.C.: U.S. Government Printing Office, 1960.

Water Resources Data for New Mexico 1965. Part 1: Surface Water Records. Washington, D.C.: U.S. Geologic Survey, 1965.

Water Resources Data for Texas 1965. Part 1: Surface Water Records. Washington, D.C.: U.S. Geologic Survey, 1965.

White, John H. *Borger, Texas: An History of the Real Facts about the Most Talked of Town in Texas and the Southwest*. 1930; reprint, Waco: Texian Press, 1974.

Williams, John C. "Testimony of John C. Williams, General Manager, Canadian River Municipal Water Authority before Water and Power Resources Subcommittee House Resources Committee July 29, 1997, <http://natural-resources.house.gov/uploadedfiles/john_c._williams_testimony_7.29.97.pdf> [Accessed May 13, 2014].

Woods, J. O. "Report on Trip to Northeastern New Mexico to Investigate Wind Erosion." In U.S/ Soil Conservation Commission Records 1919–1953, Center for Southwest Research, University of New Mexico, Albuquerque, New Mexico.

Yeo, Herbert. *Ninth Biennial Report State Engineer of New Mexico, 1928–1930*. Santa Fe: State Engineer's Office, 1930.

Secondary Sources

ARTICLES AND CHAPTERS

Allen, R. G., H. W. Price, and E. V. Reinbold, "The History, Use and Manufacture of Carbon Black." *Panhandle Plains Historical Review* 12 (1939): 24-47.

Archambeau, Ernest R. "Panhandle Pioneer Settler Recalls Origin, Early

Days of 'Old Tascosa.'" *Amarillo Times,* February 28, 1946.

Balling, Robert C. and Stephen G. Wells. "Historical Rainfall Patterns and Arroyo Activity within Zuni River Drainage Basin, New Mexico." *Annals of the Association of American Geographers* 80 (December 1990): 603–617.

Bartlett, N. D. "Discovery of the Panhandle Oil and Gas Field." *Panhandle Plains Historical Review* 12 (1939): 48–54.

Beavis, W. D., J. C. Owens, T. S. Bellows, J. A. Ludwig, and E. W. Huddleston. "Density and Developmental Stage of Range Caterpillar *Hermileuca oliviae* Cockerell, as Affected by Topographic Position." *Journal of Range Management* 34 (September 1981): 389–392.

Carlson, Paul H. "Indian Agriculture, Changing Subsistence Patterns and the Environment of the Southern Great Plains." *Journal of Agricultural History* 66 (Spring 1992): 52–60.

Fisher, C. F. "The Mesquite Problem in the Southwest." *Journal of Range Management* 3 (January 1950): 60–70.

Flores, Dan. "All the Pretty Horses: The Horse Trade and the Early American West, 1775–1825." *Montana: The Magazine of Western History* 58 (Summer 2008): 3–21.

Frankenfield, H. C. "Rivers and Floods," *Monthly Weather Review* 51 (November 1923): 604.

Grantham, V. O. "Innovations in Aqueduct Construction on the Canadian River Project." *Reclamation Age* 50 (August 1964): 53–55.

Gustavson, Thomas C. "Structural Control of Major Drainage Elements Surrounding the Southern High Plains." In *Geology and Geohydrology of the Palo Duro Basin, Texas Panhandle: A Report on the Progress of Nuclear Waste Soil Feasibility Studies 1980,* ed. Thomas C. Gustavson. Austin: University of Texas Press for Texas Bureau of Economic Geology, 1981, 129–133.

Gustavson, T. C., Robert C. Finley, and Robert W. Baumgardner Jr. "Preliminary Rates of Slope Retreat and Salt Dissolution along the Eastern Caprock Escarpment of the Southern High Plains and in the Canadian River Valley." In *Geology and Geohydrology of the Palo Duro Basin, Texas Panhandle: A Report on the Progress of Nuclear Waste Isolation Feasibility Studies 1979,* ed. T. C. Gustavson et al. Austin: University of Texas Press for the Bureau of Economic Geology, 1980, 76–81.

Hagy, Lawrence R. "History of Development of General Geology of the Panhandle Field of Texas." *Panhandle Plains Historical Review* 12 (1939): 55–63.

Hawkins, Rusty. "John C. Williams and West Texas Reclamation." *Panhandle Plains Historical Review* 77 (2004): 35–46.

Holden, William C. "The Problem of Maintaining the Solid Range on the Spur Ranch." *Southwestern Historical Quarterly* 34 (July 1930): 1–19.

Hunt, Adrian. "Conchas and Ute Reservoirs and Water Issues in Eastern New Mexico." In *Geology of the Llano Estacado,* ed. Spencer D. Lucas and Dana S. Ulmer-Scholler. Albuquerque: New Mexico Geologic Society, 2001. 48–51.

Kroch, Carol Glaubman. "Environmental Law—Water Pollution Remedies—

Application for Federal Common Law of Public Nuisance to Intrastate Stream Pollution—Committee for Consideration of Jones Falls Sewage System v. Train." *Boston College Industrial and Commercial Law Review* 18, No. 5 (1977): 929–955.

MacLish, Archibald. "The Grasslands." *Fortune,* November 1935, 58–67, 186–203.

Malin, James C. "Dust Storms—Part I." *Kansas Historical Quarterly* 14 (May 1946): 129–133.

———. "Dust Storms—Part II." *Kansas Historical Quarterly* 14 (August 1946), 265–296.

———. "Dust Storms—Part III." *Kansas Historical Quarterly* 14 (November 1946): 391–413.

Mann, Daniel H., and David J. Meltzer. "Millennial-Scale Dynamics of Valley Fills over the Past 12,000 14C Years in Northeastern New Mexico, USA." *GSA Bulletin* 119 (November–December 2007): 1433–1448.

Martin, Charles W., and William C. Johnson, "Historical Channel Narrowing and Riparian Vegetation Expansion in the Medicine Lodge River Basin, Kansas, 1871–1983." *Annals of the Association of American Geographers* 77, no. 3 (1987), 436–449.

McDuff, Cassie. "Inland Cities Could Learn from Santa Fe, N.M.'s Water Restrictions." *Riverside (California) Press-Enterprise*, Oct. 22, 2007.

Moser, T. H. "What Has Been Done about Salt Cedar at Caballo Reservoir." *Reclamation Era* 46 (May 1960): 37–40.

Mutz, Philip B. "Reservoir Storage Development in the Canadian River Basin and Related Litigation." In *Water Issues of Eastern New Mexico: Get Your Water Kicks on Route 66.* New Mexico Water Resources Research Institute Conference, Las Cruces, New Mexico, New Mexico State University, Oct. 1997, 11–15.

Nall, Gary L. "Dust Bowl Days: Panhandle Farming in the 1930s." *Panhandle Plains Historical Review* 48 (1975): 42–63.

———. "Panhandle Farming in the 'Golden Age' of American Agriculture." *Panhandle Plains Historical Review* 46 (1973): 68–93.

———. "Specialization and Expansion: Panhandle Farming in the 1920s." *Panhandle Plains Historical Review* 47 (1974): 46–67.

O'Neill, Robert V. "Is It Time to Bury the Ecosystem Concept? (With Full Military Honors, of Course!)" *Ecology* 82, no. 12 (2001): 3275–3284.

Phillips, Sarah. "FDR, Hoover and the New Rural Conservation, 1920–1932." In *FDR and the Environment*, ed. Henry L. Henderson and David B. Woolner. New York: Palgrave Macmillan, 2005. 106–138.

Presley, Mark W. "Salt Deposition Systems: An Example from the Tubbs Formation." In *Geology and Geohydrology of the Palo Duro Basin, Texas Panhandle: A Report on the Progress of Nuclear Waste Soil Feasibility Studies 1979*, ed. Thomas C. Gustavson. Austin: University of Texas Press for Texas Bureau of Economic Geology, 1980, 24–28.

Riddle, Roy. "Casimero Romero as Benevolent Don in Brief Pastoral Era." *Amarillo Times/ Amarillo Globe*, Aug. 14, 1938.

Sherow, James E. "Workings of the Geodialectic: High Plains Indians and Their Horses in the Region of the Arkansas River Valley, 1800–1870."in *A Sense of the American West: An Anthology of Environmental History*, ed. James E. Sherow (Albuquerque: University of New Mexico Press,1998), 91–112.

Sikes, Samuel and Jackie Smith, "A Vegetational Study of the Canadian River Breaks," in *The Canadian River Breaks: A Natural Area Survey* (Austin: University of Texas Press, 1975), 46–54. Available at: http://www.lib.utexas.edu/books/landscapes/publications/txu-oclc-3885697/txu-oclc-3885697.pdf [Accessed June 21, 2014]

Smith, Joe H. "The Plains Plazas: A Brief Pause in the Span of Time." *Southwest Heritage* 3, No. 2 (1969): 13–18.

Stahle, David W., and Malcom K. Cleveland. "Texas Drought History Reconstructed and Analyzed from 1698 and 1980." *Journal of Climate* 1 (January 1994): 59–74.

Stegner, Wallace. "Myths of the Western Dam," *Saturday Review*, October 23, 1965, 29–31.

Stiles, Arthur. "The Gradient Boundary—The Line between Texas and Oklahoma along the Red River." *Texas Law Review* 30 (1952): 305–323.

Trimble, Stanley W., and Alexandra C. Mendel. "The Cow as a Geomorphic Agent: A Critical Review." *Geomorphology* 13, no. 1 (1995): 233–323.

United States Weather Bureau, "Flood on the South Canadian River in Oklahoma and Indian Territory, October 1–4, 1904," *Monthly Weather Review* 32 (November 1904): 522.

VanLooy, Jeffrey, and Charles W. Martin. "Channel and Vegetation Change on the Cimarron River, Southwestern Kansas, 1953–2001." *Annals of the Association of American Geographers* 95, No. 4 (2005): 727-739.

Vigness, Winifred. "Municipal Government in Lubbock." In *A History of Lubbock*, ed. Lawrence L. Graves. Lubbock: West Texas Museum Association, 1962. 335–384.

Williams, John C. "Eleven Thirsty Cities Seeking Water Now and into the Future" *Panhandle Plains Historical Review* 78 (2004). 48–57.

Wisniewski, P. S., and F. J. Passaglia. "Epirogenic Controls on Canadian River Incision and Landscape Evolution, Great Plains of Northern New Mexico." *Journal of Geology* 110 (July 2002): 437–456.

Woods, Winton D., Jr., and Kenneth R. Reed. "The Supreme Court and Interstate Environmental Quality: Some Notes on the *Wyandotte* Case." *Arizona Law Review* 12 (1970): 691–715.

Zoch, Richmond T. "Rivers and Floods," *Monthly Weather Review* 63 (May 1935): 170–172.

Zurer, Rachel. "The Birds and the Beetle(s)." *High Country News*, October 1, 2010.

Secondary Sources: Dissertations and Theses

Bickers, Margaret A. "Three Cultures, Four Hooves and One River: The Canadian River in New Mexico and Texas, 1848–1938." Ph.D. diss., Kansas State University, 2010.

Cornebise, Alfred Emile. "A Decade in the History of the Matador Land and Cattle Company, Ltd., 1919–1928." Master's thesis, Texas Technological College, 1958.

Ethridge, Devine. "The Role of the North American Monsoon in the Landscape Evolution of the Southwest United States." Master's thesis, University of New Mexico, 2000.

Flynn, Sean J. "Living History: John L. McCarty and the Texas Panhandle." Ph.D. diss., Texas Tech University, 1999.

Hatcher, Averlyne M. "The Water Problem of the Matador Ranch." Master's thesis, Texas Technological College, 1944.

Hill, David P. "Mammals of the Lake Meredith Recreation Area and Alibates National Monument." Master's thesis, West Texas State University, 1984.

Hofma, Georgie E. "Lake Meredith Water Quality Case Study: Economic Evaluations." Master's thesis, West Texas State University, 1984.

Kelly, Charles Robert. "The Canadian River Municipal Water Authority Project in West Texas: A Geographic Analysis." Ph.D. diss., University of Oklahoma, 1971.

Kessler, L. Gifford, II. "Channel Sequences and Braided Stream Development in the South Canadian River, Hutcheson, Roberts and Hemphill Counties, Texas." Ph.D. diss., University of New Mexico, 1972.

Web-Sites

"About Pantex," <http://www.pantex.com/about/pages/history.aspx> [Accessed May14, 2014].

"Amarillo Area Foundation: Hospital District." <http://www.amarilloarea-foundation.org/page.aspx?pid=259> [Accessed May 14, 2014].

"Amarillo Hospital District," <http://www.amarillohospitaldistrict.org> [Accessed April 22, 2012].

"Amarillo Population Growth," <http://www.goldenspread.us/Amarillo/history?timeline.html> [Accessed Apr. 22, 2012].

"Amarillo, TX: Yearly Precipitation Totals 1892–2008," <http://www.srh.noaa.gov/ama/?n=yearly_precip> [Accessed May 14, 2014].

Anderson, H. Allen. "Borger, TX," *The Handbook of Texas Online,* <http://www.tshaonline.org/handbook/online/articles/heb10> [Accessed May 17, 2014].

BBER-University of New Mexico. "Historical Census Data," < http://bber.unm.edu/census/cenhist.htm> [Accessed May 14, 2014].

Bowser, David. "Protestors Demand Rehearing on Canadian River Deci-

sion," <http://www.livestockweekly.com/papers/99/11/04/whlcanadian.
asp>[Accessed May 14, 2014].

"Canadian River Boundary Dispute Enters Round Two," <http://www.live-
stockweekly.com/papers/96/12/19/index.html> [Accessed April 22, 2012].

"Canadian River Project." <http://www.usbr.gov/projects/Project.
jsp?proj_Name=Canadian%20River%20Project&pageType=ProjectHist
oryPage> and <http://www.usbr.gov/projects//ImageServer?imgName=
Doc_1303158200779.pdf> [Accessed May 15, 2014].

"Cannon, Clarence Andrew." *Biographical Directory of the United States Congress,* <http://bioguide.congress.gov/scripts/biodisplay.pl?index=C000117>
[Accessed May 15, 2014].

"A Celebration of New Mexico's First Women Lawyers," <http://www.nmbar.
org/Attorneys/BarsLegalGroups/NMWomensBar/nmwomenbarfirstfema-
leattys.html> and <http://www.nmbar.org/Attorneys/BarsLegalGroups/
NMWomensBar/womanbarfirst100.pdf> [Accessed May 15, 2014].

"Cold and Warm Episodes by Season," <http://www.cpc.ncep.noaa.gov/prod-
ucts/analysis_monitoring/ensostuff/ensoyears.shtml> [Accessed May 15,
2014].

"Cotton Marketing," <http://www.oldandsold.com/articles04/textiles5.shtml>
[Accessed May 15, 2014].

"History of CRMWA," <Http://crmwa.com/history-of-crmwa/> [Accessed
May 15, 2014].

"Cuyahoga River Fire," <http://www.ohiohistorycentral.org/entry.
php?rec=1642> [Accessed May 15, 2014].

"Desalination of Brackish Water and Seawater." <http://www.prominent.com/
desktopdefault.aspx/tabid-190/570_read-2273/> [Accessed May 15, 2014].

"Digest of Federal Resource Laws of Interest to the US Fish and Wildlife
Service: Federal Water Pollution Control Act (Clean Water Act)," <http://
www.fws.gov/laws/lawsdigest/fwatrpo.html> [Accessed May 15, 2014].

"Fishing Lake Meredith," <http://www.tpwd.state.tx.us/fishboat/fish/recre-
ational/lakes/meredith/> [Accessed May 15, 2014].

George, Ricky. "Water Pollution Expert Retiring," <http://amarillo.com/sto-
ries/2001/06/01/new_expert.shtml> [Accessed May 15, 2014].

Horton, Jerome S. "The Development and Perpetuation of the Permanent
Tamarisk Type in the Phreatophyte Zone of the Southwest," 124–127. U.S.
Department of Agriculture, Forest Service Research, Rocky Mountain
Research Station, Southwest Watershed Science Team, >http://www.rmrs.
nau.edu/awa/ripthreatbib/horton_devperptamarisk.pdf> [Accessed 14 May,
2014].

"Lake Alan Henry Water Supply Project," <http://www.team-psc.com/siwa_
lah.html> [Accessed May 15, 2014].

"Lubbock Precipitation 1911–Present," <http://www.srh.noaa.gov/
lub/?n=climate-klbb-pcpn> [Accessed May 15, 2014].

"Lubbock Tornado," <http://www.lubbockonline.com/lubbocktornado.shtml>

[Accessed May 15, 2014].

Odintz, Mark. "Roberts County," *The Handbook of Texas Online*, <http://www.tshaonline.org/handbook/online/articles/hcr08> [Accessed May 15, 2014].

"Office of the Delaware River Master," <http://water.usgs.gov/osw/odrm/decree.html> [Accessed May 15, 2014].

Oklahoma v. Texas 260 U.S. 606 (1923), <http://supreme.justia.com/cases/federal/us/260/606/> [Accessed May 17, 2014].

"Our History,"<http://www.aaf-hf.org/about/history.htm> [Accessed April 22, 2012].

"An Outline History of the Arkansas River," http://www.tulsaweb.com/port/history.htm [Accessed April 24, 2012].

Phillips, Wes. "Fish of Lake Meredith," <http://www.panhandlenation.com/geographica/fish/fish.html> [Accessed Apr. 24, 2012].

"Population History of Selected Cities, 1850–2000,"<http://www.texasalmanac.com/topics/population> [Accessed May 15, 2014].

Powell, Michael V. "Riparian Boundaries in Texas," <http://www.lockelord.com/~/media/Files/NewsandEvents/News/2006/03/Riparian%20Boundaries%20in%20Texas/Files/Riparian/FileAttachment/Riparian%20Boundaries%20-%20Powell.pdf> [Accessed May 15, 2014].

Price, B. Byron. "Haley, James Evetts, Sr.," *The Handbook of Texas Online*. <http://www.tshaonline.org/handbook/online/articles/fhahj> [Accessed May 15, 2014].

"Reservoir Terminology." <http://midgewater.twdb.state.tx.us/swrweb/swr/hydro> [Accessed April 24, 2012].

Sheets, Jody. "A Brief Commentary on the Abuse of Sovereign Immunity by the State of Texas along Texas Rivers," <http://www.texas-wildlife.org/index.php?option=com_content&view=article&id=44:a-brief-commentary-on-the-abuse-of-sovereign-immunity-by-the-state-of-texas-along-texas-rivers&catid=16:water&itemid=> [Accessed January 28, 2014].

"Sierra Club Timeline—History,"<http://www.sierraclub.org/history/timeline.aspx> [Accessed May 15, 2014].

Stene, Eric A. "The Canadian River Project." <http://www.usbr.gov/projects//ImageServer?imgName=Doc_1303158200779.pdf> [Accessed May 15, 2014].

"The Taylor Grazing Act,." <http://www.blm.gov/wy/st/en/field_offices/Casper/range/taylor.1.html> [Accessed May 15, 2014].

"Toxaphene," <http://www.epa.gov/pbt/pubs/toxaphene.htm> [Accessed May 15, 2014].

"Ute Lake State Park," <http://geoinfo.nmt.edu/tour/state/ute_lake/home.html> [Accessed May 15, 2014].

"Water Use in Santa Fe 2009," <http://www.santafenm.gov/how_much_water_do_we_use_reports_and_studies> Accessed [May 15, 2014].

"Water Restrictions in New Mexico Help Shore Up Dwindling Supplies.," <http://www.uswaternews.com/archives/arcconserv/6newmex.html> [Accessed April 22, 2012].

"What Is Meant by the Term Drought?," <http://www.wrh.noaa.gov/fgz/science/drought.php?wfo=fgz> [Accessed May 15, 2014].

"Worley, Francis Eugene," *Biographical Directory of the United States Congress,* <http://bioguide.congress.gov/scripts/biodisplay.pl?index=W000744> [Accessed May 15, 2014].

Secondary Sources: Brochures and Pamphlets

Bell, Ann E., and Shelly Morrison. *Analytical Study of the Ogallala Aquifer in Lubbock County, Texas.* Austin: Texas Department of Water Resources, 1978.

Berkstrasser, Charles F., and Walter A. Mourant. *Groundwater Resources and Geology of Quay County, New Mexico.* Socorro: New Mexico State Bureau of Mines and Mineral Resources, 1960.

Bovey, R. W., and R. E. Meyer. *The Response of Honey Mesquite to Herbicides.* Texas A&M Bulletin B-1363, College Station: Texas A&M University Press, 1981.

Canadian Breaks: A Natural Area Survey. Part 7 of 8. Austin: University of Texas Press, 1975.

Canadian River Breaks: A Natural Area Survey. Part 7 of 7. Austin: University of Texas Press, 1975.

Chudnoff, Mustafa, and Linda Logan. *Groundwater Relationship between New Mexico and Texas along the State Line in the Southern High Plains.* SPGH-95-01 Hydrology Section, Special Projects Division. Santa Fe: New Mexico State Engineer's Office, 1995.

Dollahite, J. W., and W. V. Anthony. *Nutrition in Cattle on an Unbalanced Diet of Mesquite Beans.* Agricultural Experiment Station Progress Report # 1931. College Station: Texas A&M University, 1957.

Dolliver, Paul N. *Cenozoic Evolution of the Canadian River Basin.* Baylor Geological Studies Bulletin No. 42. Waco, Tex.: Baylor University Press, 1984.

Leggat, E. R. *Geology and Groundwater Resources of Lynn County, Texas.* Austin: Texas Board of Water Engineers, 1952.

Mahler, B. J., and P. C. Van Metre. *Effects of Oil and Gas Production on Lake Meredith Sediments, 1964–1999.* USGS Fact Sheet 072-01. Washington, DC: U.S. Geological Survey, 2001.

McGinty, Allan, and Darrell Ueckert. *Brush Busters: Hints to Beat Mesquite.* Agricultural Experiment Station Report 2005 L-5144. College Station: Texas A&M University, 2005.

New Mexico State Engineer's Office. *Preliminary Report on the Geology of the Ute Dam Site, Quay County, New Mexico.* Vol. 3: *Canadian River Storage Sites Investigating Ute Reservoir.* Santa Fe: State Engineer's Office Technical Division, 1961.

Shipley, John, and Cecil Rieger. *Winter Wheat Grazing and Irrigation Water Management Northern High Plains.* Texas A&M Agricultural Experiment Station Progress Report PR 3030. College Station: Texas A&M Agricultural Experiment Station, 1972.

Schumm, S. A., and R. W. Lichty. *Channel Widening and Flood-Plain Con-*

struction along Cimarron River in Southwestern Kansas. Geological Survey Professional Paper 352-D. Washington D.C.: U.S. Government Printing Office, 1963.

Spiegel, Zane. *Natural Discharge of Saline Water from Permian Rocks in the Basins of the Canadian, Red, Brazos and Colorado Rivers in New Mexico, Texas and Oklahoma.* Santa Fe: New Mexico Office of State Engineer, 1966.

Texas Agricultural Experiment Station. *Mesquite: Growth, Development, Management, Economics, Uses.* College Station: Texas A & M University, 1973.

Ueckert, Darrell, and Allan McGinty. *Brush Busters: How to Estimate Cost for Controlling Small Mesquite.* Texas A&M Agricultural Experiment Station Bulletin L-5291. College Station: Texas A&M University Press, 1999.

———. *Brush Busters: How to Estimate Cost for Controlling Small Cedars.* Texas A&M Agricultural Experiment Station Bulletin L-5292. College Station: Texas A&M University Press, 1999.

Unger, Paul W. *Dryland Winter Wheat and Grain Sorghum Cropping Systems Northern High Plains of Texas.* Texas A&M Agricultural Extension Bulletin B-1126. College Station: Texas A&M University Press, 1972.

Warren, J. A. *Agriculture in the Central Part of the Semiarid Portion of the Great Plains.* USDA Bulletin of Plant Industry, no. 215 .Washington D.C.: United States Government Publishing Office, 1911.

Webb, Walter P. *More Water for Texas: The Problem and the Plan.* Austin: University of Texas Press, 1954.

Secondary Sources: Monographs

Abbe, Donald, Paul H. Carlson, and David J. Murrah. *Lubbock and the South Plains: An Illustrated History.* Chatsworth, Calif.: Windsor Publications, 1989.

Amarillo Genealogy Society. *Texas Panhandle Forefathers,* comp. Barbara C. Spray. Amarillo: Amarillo Genealogy Society, 1983.

Anderson, Gary Clayton. *The Indian Southwest 1580–1830: Ethnogenesis and Reinvention.* Norman: University of Oklahoma Press, 1999.

Archer, Jeanne S., and Stephanie Kadel Taras. *Touching Lives: The Lasting Legacy of the Bivins Family.* Amarillo: Tell Studios, 2009.

Bamforth, Douglas B. *Ecology and Human Organization on the Great Plains.* New York: Plenum Press, 1988.

Berry, John R. *Rising Tide: The Great Mississippi Flood of 1927 and How It Changed America.* New York: Simon and Schuster, 1997.

Bonnifield, Paul. *The Dust Bowl: Men, Dirt and Depression.* Albuquerque: University of New Mexico Press, 1979.

Brooks, James R. *Captives and Cousins: Slavery, Kinship and Community in the Southwest Borderlands.* Chapel Hill: University of North Carolina Press, 2002.

Brooks, Karl Boyd. *Before Earth Day: The Origins of American Environmental*

Law, 1945–1970. Lawrence: University Press of Kansas, 2009.

Brune, Gunnar. *Springs of Texas.* 2nd Ed. College Station: Texas A&M University Press, 2002.

Burges, Austin Earles. *Soil Erosion Control: A Practical Exposition of the New Science of Soil Conservation for Students, Farmers and the General Public.* Atlanta: Turner E. Smith and Co., 1936.

Campbell, Linda. *Endangered and Threatened Animals of Texas: Their Life History and Management.* Austin: Texas Parks and Wildlife Resource Protection Division, 1995.

Carlson, Paul H. *Amarillo: The Story of a Western Town.* Lubbock: Texas Tech University Press, 2006.

———. *Deep Time and the Texas High Plains: History and Geology.* Lubbock: Texas Tech University Press, 2005.

———. *Empire Builder in the Texas Panhandle: William Henry Bush.* College Station: Texas A&M University Press, 1996.

Clark, Ira G. *Water in New Mexico: A History of Its Management and Use.* Albuquerque: University of New Mexico Press, 1987.

Clements, Frederick E. *Plant Succession: An Analysis of the Development of Vegetation.* Carnegie Institute of Washington Publication 242. Washington, D.C.: Gibson Brothers Press, 1916.

Cox, Mary L. *History of Hale County, Texas.* Plainview, Tex.: 1937.

Cunfer, Geoff. *On The Great Plains.* College Station: Texas A&M University Press, 2005.

Cypher, John. *Bob Kleberg and the King Ranch: A Worldwide Sea of Grass.* Austin: University of Texas Press, 1996.

Dawson County Historical Commission. *Dawson County History.* Lubbock: Craftsman Printers and Taylor Publishing Co., 1981.

Dobkins, Betty Eakle. *The Spanish Element in Texas Water Law.* Austin: University of Texas Press, 1959.

Dunn, Thomas, and Luna B. Leopold. *Water in Environmental Planning.* New York: W. H. Freeman and Company, 1978.

Egan, Timothy. *The Worst Hard Time: The Untold Story of Those Who Survived the Great American Dust Bowl.* Boston: Houghton Mifflin Co., 2006.

El-Ashy, Mohamed T., and Diana C. Gibbons (eds.). *Water and Arid Lands of the Western United States.* New York: Cambridge University Press, 1988.

Erickson, John. *The Modern Cowboy.* Lincoln: University of Nebraska Press, 1981.

———. *Through Time and the Valley.* Denton: University of North Texas Press, 1995.

Fitzgerald, Deborah. *Every Farm a Factory: The Industrial Ideal in American Agriculture.* New Haven, Conn.: Yale University Press, 2003.

Flores, Dan. *The Natural West: Environmental History in the Great Plains and Rocky Mountains.* Norman: University of Oklahoma Press, 2001.

Frink, Maurice, W. Turrentine Jackson, and Agnes Wright Spring. *When Grass*

Was King: Contributions to the Western Range Cattle Industry Study. Boulder: University of Colorado Press, 1956.

Gates, Paul Wallace. *The History of Land Law Development.* Washington, D.C.: U.S. Government Printing Office, 1968; reprint, New York: Arno Press, 1979.

Gilliland, Charles E. *Endangered Species Act: Impact on Texas Rural Land Values.* College Station: Texas Real Estate Center, 1995.

Gracy, David B., II. *Littlefield Lands: Colonization on the Texas Plains, 1912–1920.* Austin: University of Texas Press, 1968.

Graf, William L. *Wilderness Preservation and the Sagebrush Revolution.* Savage, Md.: Rowman and Littlefield Publishers, 1990.

Graves, Lawrence L. (ed.). *A History of Lubbock.* Lubbock: West Texas Museum Association, 1962.

Green, Donald E. *Land of the Underground Rain: Irrigation on the Texas High Plains, 1910–1970.* Austin: University of Texas Press, 1973.

Griggs, William C. (ed.). *A Pictorial History of Lubbock, Texas 1880–1950.* Lubbock: Lubbock County Historical Commission, 1976.

Gustavson, Thomas C., et al. (eds.). *Geology and Geohydrology of the Palo Duro Basin, Texas Panhandle: A Report on the Progress of Nuclear Waste Soil Feasibility Studies, 1979.* Austin: University of Texas for the Bureau of Economic Geology, 1980.

———. *Geology and Geohydrology of the Palo Duro Basin, Texas Panhandle: A Report on the Progress of Nuclear Waste Soil Feasibility Studies, 1980.* Austin: University of Texas for the Bureau of Economic Geology, 1981.

Haley, J. Evetts. *Charles Goodnight: Cowman and Plainsman.* Norman: University of Oklahoma Press, 1949.

———. *Life on the Texas Plains: Photographs of E. Erwin Smith.* Austin: University of Texas Press, 1952.

———. *The XIT Ranch of Texas and the Early Days of the Llano Estacado.* New ed. Norman: University of Oklahoma Press, 1953.

Hall, G. Emlen. *Four Leagues of Pecos: A Legal History of the Pecos Grant, 1880–1933.* Albuquerque: University of New Mexico Press, 1984.

———. *High and Dry: The Texas–New Mexico Struggle for the Pecos River.* Albuquerque: University of New Mexico Press, 2002.

Hämäläinen, Pekka. *The Comanche Empire.* New Haven, Conn.: Yale University Press, 2008.

Hays, Samuel P. *Conservation and the Gospel of Efficiency: The Progressive Conservation Movement, 1890–1920.* Cambridge, Mass.: Harvard University Press, 1959.

Hays, Samuel P., and Barbara D. Hays. *Beauty, Health and Permanence: Environmental Politics in the United States, 1955–1985.* New York: Cambridge University Press, 1987.

Henderson, Henry L., and David B. Woolner (eds.). *FDR and the Environment.* New York: Palgrave Macmillan, 2005.

Hill, Frank P., and Pat Hill Jacobs. *Grassroots Upside Down: A History of Lynn County, Texas*. Austin: Nortex Press, 1986.

Holliday, Vance T. *Paleoindian Geoarchaeology of the Southern High Plains*. Austin: University of Texas Press, 1997.

Hoyt, William G., and Walter B. Langbein. *Floods*. Princeton, N.J.: Princeton University Press, 1955.

Hundley, Norris, Jr. *The Great Thirst: Californians and Water 1770s–1990s*. Berkeley: University of California Press, 1992.

———. *Water and the West: The Colorado River and the Politics of Water in the American West*. 2nd ed. Berkeley: University of California Press, 2009.

Hurt, R. Douglas. *American Agriculture: A Brief History*. Ames: Iowa State University Press, 1994.

———. *Problems of Plenty: The American Farmer in the Twentieth Century*. Chicago: Ivan R. Dee, 2002.

Hutchins, Wells A. *The Texas Law of Water Rights*. Austin: Texas Board of Water Engineers, 1961.

———. *Water Rights Laws in the Nineteen Western States* Vol. 2. Washington, D.C.: U.S. Government Printing Office, 1974.

Hutchinson County Historical Commission. *History of Hutchinson County, Texas*. Dallas: Taylor Press, 1980.

Irwin, Douglas A. *Peddling Protectionism: Smoot-Hawley and the Great Depression*. Princeton, N.J.: Princeton University Press, 2011.

Jackson, Donald C. *Building the Ultimate Dam: John S. Eastwood and the Control of Water in the West*. Norman: University of Oklahoma Press, 2005.

John, Elizabeth A. H. *Storms Brewed in Other Men's Worlds: The Confrontation of Indians, Spanish, and French in the Southwest, 1540–1795*. 2nd ed. Norman: University of Oklahoma Press, 1996.

Johnson, Vance. *Heaven's Tableland: The Dust Bowl Story*. 1947; reprint, New York: De Capo Press, 1974.

Leopold, Luna B. *Water: A Primer*. San Francisco: W. H. Freeman and Co., 1974.

———. *A View of the River*. Cambridge, Mass.: Harvard University Press, 1994.

Lintz, Christopher Ray. *Architecture and Community Variability within the Antelope Creek Phase of the Texas Panhandle*. Studies on Oklahoma's Past No. 14. Norman: University of Oklahoma Press, 1986.

Littlefield, Douglas R. *Conflict on the Rio Grande: Water and the Law, 1879–1939*. Norman: University of Oklahoma Press, 2008.

Lucas, Spencer D., and Dana S. Ulmer-Scholler (eds.). *Geology of the Llano Estacado*. Albuquerque: New Mexico Geologic Society, 2001.

Malin, James C. *Winter Wheat in the Golden Belt of Kansas: A Study in Adaptation to Subhumid Geographical Environment*. 1944; reprint, New York: Octagon Books, 1973.

Mann, Charles C., and Mark L. Plumber. *Noah's Choice: The Future of Endangered Species.* New York: Alfred A. Knopf, 1995.

McCarty, John L. *Maverick Town: The Story of Old Tascosa.* Norman: University of Oklahoma Press, 1946.

McCray, Arthur W., and Frank W. Cole. *Oil Well Drilling Technology.* Norman: University of Oklahoma Press, 1959.

McCully, Patrick. *Silenced Rivers: The Ecology and Politics of Large Dams.* Enlarged and updated ed. New York: Zed Books, 2001.

McHugh, Tom. *Time of the Buffalo.* Lincoln: University of Nebraska Press, 1979.

Melosi, Martin. *The Sanitary City: Urban Infrastructure in America from Colonial Times to Present.* Baltimore: Johns Hopkins University Press, 2000.

Mesquite: Growth and Development, Management, Economics, Control, Uses. College Station: Texas A & M University, Texas Agricultural Experiment Station, 1973.

Miller, Thomas Lloyd. *The Public Lands of Texas 1519–1970.* Norman: University of Oklahoma Press, 1972.

Moore, Richard R. *West Texas after the Discovery of Oil: A Modern Frontier.* Austin: Pemberton Press, 1971.

Morgan, R. P. C. *Soil Erosion and Conservation.* 3rd ed. Malden, Mass.: Blackwell Publishing, 2005.

Morrow, Baker H. *A Dictionary of Landscape Architecture.* Albuquerque: University of New Mexico Press, 1987.

Muehlberger, William R., Sally J. Muehlberger, and L. Greer Price. *High Plains of Northeastern New Mexico: A Guide to Geology and Culture.* Socorro: New Mexico Bureau of Geology and Mineral Resources of the New Mexico Institute of Mining and Technology, 2005.

Myres, Samuel D. *The Permian Basin: Petroleum Empire of the Southwest.* Vol. 2: *Era of Advancement.* El Paso: Permian Press, 1977.

Nash, Roderick. *Wilderness and the American Mind.* 4th ed. New Haven, Conn.: Yale University Press, 2001.

Nelson, Robert H. *Public Lands and Private Rights: The Failure of Scientific Management.* Savage, Md.: Rowman and Littlefield, 1995.

Nostrand, Richard L. *The Hispano Homeland.* Norman: University of Oklahoma Press, 1992.

Oldham County Historical Commission. *Oldham County 1881–1981.* Lubbock: Craftsman Printers, 1981.

Opie, John. *Ogallala: Water for a Dry Land.* 2nd ed. Lincoln: University of Nebraska Press, 2000.

Pearce, William M. *The Matador Land and Cattle Company.* Norman: University of Oklahoma Press, 1964.

Pearson, Byron E. *Still the Wild River Runs: Congress, the Sierra Club and the Fight to Save Grand Canyon.* Tucson: University of Arizona Press, 2002.

Peffer, E. Louise. *The Closing of the Public Domain: Disposal and Reservation Policies, 1900–1950.* New York: Arno Press, 1972.

Pindar, George F., and Michael A. Celia. *Subsurface Hydrology.* Hoboken, N.J.: John Wiley and Sons, 2006.

Pisani, Donald. *Water and American Government: The Reclamation Bureau, National Water Policy and the West, 1902–1935.* Berkeley: University of California Press, 2002.

———. *Water, Land and Law in the West: The Limits of Public Policy, 1850–1920.* Lawrence: University of Kansas Press, 1996.

Price, B. Byron, and Frederick Rathjen. *The Golden Spread: An Illustrated History of Amarillo and the Texas Panhandle.* Northridge, Calif.: Windsor Publications, 1986.

Rathjen, Frederick W. *The Texas Panhandle Frontier.* Rev.ed. with introduction by Elmer Kelton. Lubbock: Texas Tech University Press, 1996.

Reeves, C. C., Jr., and Judy A. Reeves. *The Ogallala Aquifer of the Southern High Plains.* Vol. 1: *Geology.* Lubbock: Estacado Books, 1996.

Reisner, Marc. *Cadillac Desert: The American West and Its Disappearing Water.* Rev. and updated ed. New York: Penguin Books, 1993.

Remley, David. *Bell Ranch: Cattle Ranching in the Southwest, 1824–1947.* Rev.ed. Las Cruces, N.M.: Yucca Tree Press, 2000.

Riney-Kehrberg, Pamela. *Rooted in Dust: Surviving Drought and Depression in Southwestern Kansas.* Lawrence: University Press of Kansas, 1994.

Rister, Carl Coke. *Oil! Titan of the Southwest.* With a foreword by E. DeGolyer. Norman: University of Oklahoma Press, 1949.

Robinson, Pauline Durrett, and R. L. Robinson. *Cowman's Country: Fifty Frontier Ranches in the Texas Panhandle 1876–1887.* Amarillo, Tex.: Paramount Publishing Company, 1981.

Rogers, Ken E. *The Magnificent Mesquite.* Austin: University of Texas Press, 2000.

Rome, Adam. *The Bulldozer in the Countryside: Suburban Sprawl and the Rise of American Environmentalism.* New York: Cambridge University Press, 2001.

Rosgen, Dave. *Applied River Morphology.* 2nd ed. Pagosa Springs, Colo.: Wildland Hydrology, 1996.

Rothman, Hal K. *Saving the Planet: The American Response to the Environment in the Twentieth Century.* Chicago: Ivan R. Dee, 2000.

Schofield, Donald F. *Indians, Cattle, Ships and Oil: The Story of W. M. D. Lee.* Austin: University of Texas Press, 1985.

Sears, Paul. *Deserts on the March.* 3rd ed. Norman: University of Oklahoma Press, 1959.

Sherk, George William. *Dividing the Waters: The Resolution of Interstate Water Conflicts in the United States.* The Hague: Kluwer Law International, 2000.

Sherow, James E. (ed.). *The Grasslands of the United States: An Environmental History.* Santa Barbara, Calif.: ABC-Clio, 2007.

———. (ed.). *A Sense of the American West*. Albuquerque: University of New Mexico Press, 1998.

———. *Watering the Valley: Development along the High Plains Arkansas River, 1870–1950*. Lawrence: University Press of Kansas, 1990.

Shlaes, Amity. *The Forgotten Man: A New History of the Great Depression*. New York: Harper Collins, 2007.

Spikes, Nellie Witt. *As a Farm Woman Thinks: Life and Land on the Texas High Plains, 1890–1960*. Ed. Geoff Cunfer, Plainsword by Sandra Scofield. Lubbock: Texas Tech University Press, 2010.

Stanley, F. *The Early Days of the Oil Industry in the Texas Panhandle*. Borger, Tex.: Hess Publishing Co., 1973.

———. *The Stinnett, Texas Story*. Nazareth, Tex.: Self-published, 1974.

Stewart, Omer C. *Forgotten Fires: Native Americans and the Transient Wilderness*. Ed. with introduction by Henry T. Lewis and M. Kat Anderson. Norman: University of Oklahoma Press, 2002.

Stradling, David (ed.). *Conservation in the Progressive Era: Classic Texts*. Foreword by William Cronon. Seattle: University of Washington Press, 2004.

Sullivan, Dulcie. *The LS Brand: The Story of a Texas Panhandle Ranch*. Austin: University of Texas Press, 1968.

Swan, Claire L. *Scottish Cowboys and Dundee Investors*. Dundee, Scotland: Aberty Historical Society, 2004.

Tarlock, Dan A., James N. Corbridge Jr., and David H. Getches. *Water Resources Management: A Casebook in Law and Public Policy*. 5th ed. New York: Foundation Press, 2002.

Terry County Historical Society. *Terry County Texas*. Clanton, Ala.: Heritage Publishing Consultants, 2002.

Thomas, Myrna Tryon. *The Windswept Land: A History of Moore County, Texas*. Dumas, Tex.: Self published, 1967.

Thomas, Ronny G. *The Geomorphic Evolution of the Pecos River System*. Baylor Geological Studies Bulletin No. 22. Waco, Tex: Baylor University, 1972.

Vallentine, John R. *Grazing Management*. San Diego, Calif.: Academic Press, 1990.

Wallace, Ernest, and E. Adamson Hoebel. *The Comanches: Lords of the South Plains*. Norman: University of Oklahoma Press, 1952.

Walters, Rupert C. S. *Dam Geology*. London: Butterworths and Co., 1962.

Warne, William E. *The Bureau of Reclamation*. New York: Praeger Publishers, 1973.

Water Issues of Eastern New Mexico: Get Your Water Kicks on Route 66. New Mexico Water Resources Research Institute Conference, Las Cruces: New Mexico Water Resources Institute,1997.

Weatherford, Guy D. *Water and Agriculture in the Western United States: Conservation, Reallocation and Markets*. In association with Lee Brown, Helen Ingram, and Dean Mann. Boulder, Colo.: Westview Press, 1982.

Weaver, Ken. *Dam Foundation Grouting.* New York: American Society of Civil Engineers, 1991.

White Deer Land Museum, Anna Davidson, and Deborah Chambers. *Pampa.* Charleston, S.C.: Arcadia Publishing, 2010.

Wilkinson, Charles E. *Crossing the Next Meridian: Land, Water and the Future of the West.* Washington, D.C.: Island Press, 1992.

Williams, Garnett P., and M. Gordon Wolman. *Downstream Effects of Dams on Alluvial Rivers.* USGS Professional Paper 1286. Washington, D.C.: U.S. Government Printing Office, 1984.

Wilson, Edmund O. *Biophilia.* Cambridge, Mass.: Harvard University Press, 1984.

Wilson, Priscilla H. *A Pioneer Love Story: The Letters of Minnie Hobart.* Shawnee Mission, Kans.: TeamTech Press, 2008.

Wood, W. Raymond (ed.). *Archaeology of the Great Plains.* Lawrence: University Press of Kansas, 1998.

Worster, Donald. *Dust Bowl: The Southern Plains in the 1930s.* 25th anniversary ed. New York: Oxford University Press, 2004.

———. *Nature's Economy: The Roots of Ecology.* San Francisco, Calif.: Sierra Club Books, 1977.

———. *Rivers of Empire: Water, Aridity, and the Growth of the American West.* New York: Oxford University Press, 1985.

Wozniak, Frank E. *Across the Caprock: A Cultural Resources Survey on the Llano Estacado and the Canadian River Valley of Eastern New Mexico for the Bravo CO2 Pipeline.* Albuquerque: Office of Contract Archaeology, University of New Mexico, 1985.

Wright, Henry A., and Arthur W. Bailey. *Fire Ecology of the United States and Southern Canada.* New York: John Wiley and Sons, 1982.

Younger, Paul L. *Groundwater in the Environment: An Introduction.* Malden, Mass.: Blackwell Publishing, 2007.

INDEX

Page numbers in italics indicate illustrations.
All listed towns are in Texas unless otherwise indicated.